Oliver Peter Gray

Sea-weeds, Shells and Fossils

Oliver Peter Gray

Sea-weeds, Shells and Fossils

ISBN/EAN: 9783744719339

Printed in Europe, USA, Canada, Australia, Japan

Cover: Foto ©berggeist007 / pixelio.de

More available books at **www.hansebooks.com**

SEA-WEEDS,

SHELLS AND FOSSILS.

BY

PETER GRAY, A.B.S. Edin.;

AND

B. B. WOODWARD,

Of the British Museum (Natural History), South Kensington.

ARDVA · QVÆ · PVLCRA

LONDON:

SWAN SONNENSCHEIN, Le BAS & LOWREY,

PATERNOSTER SQUARE.

BUTLER & TANNER,
THE SELWOOD PRINTING WORKS,
FROME, AND LONDON.

SEA-WEEDS.

By PETER GRAY.

ALGÆ, popularly known as sea-weeds, although many species
are inhabitants of fresh water, or grow on moist ground, may
be briefly described as cellular, flowerless plants, having no
proper roots, but imbibing nutriment by their whole surface
from the medium in which they grow.   As far as has been
ascertained, the total number of species is about 9000 or 10,000.
Many of them are microscopic, as the Desmids and Diatoms,
others, as Lessonia, and some of the larger Laminariæ (oarweeds),
are arborescent, covering the bed of the sea around the coast
with a submarine forest ; while in the Pacific, off the north-
western shores of America, Nereocystis, a genus allied to Laminaria, has a stem over 300 feet in length, which, although not
thicker than whipcord, is stout enough to moor a bladder, barrel-
shaped, six or seven feet long, and crowned with a tuft of fifty
leaves or more, each from thirty to forty feet in length.   This
vegetable buoy is a favourite resting place of the sea otter ; and
where the plant exists in any quantity, the surface of the sea is
rendered impassable to boats.   The stem of Macrocystis, which
"girds the globe in the southern temperate zone," is stated to
extend sometimes to the enormous length of 1500 feet.   It is
no thicker than the finger anywhere, and the upper branches
are as slender as pack-thread ; but at the base of each leaf there
is placed a buoy, in the shape of a vesicle filled with air.
   Although the worthlessness of Algæ has been proverbial, as
in the "alga inutile" of Horace and Virgil's "projecta vilior
alga," they are not without importance in botanical economics.
A dozen or more species found in the British seas are made
use of, raw or prepared in several ways, as food for man.
Of these edible Algæ, Dr. Harvey considers the two species
of Porphyra, or laver, the most valuable.   Berkeley says, " The
best way of preparing this vegetable or condiment, which

3

is extremely wholesome, is to heat it thoroughly with a little
strong gravy or broth, adding, before it is served on toast, a
small quantity of butter and lemon juice." A species of Nostoc
is largely consumed in China as an ingredient in soup. A
similar use is made of Enteromorpha intestinalis in Japan.
Many species of fish and other animals, turtle included, live upon
sea-weed. Fucus vesiculosus is a grateful food for cattle. In
Norway, cattle, horses, sheep, and pigs are largely fed upon it,

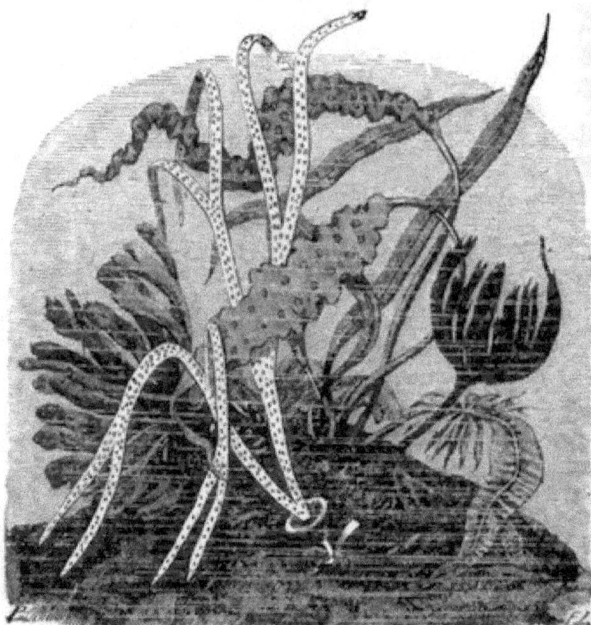

Fig. 1.   Group of Sea-weeds (chiefly Laminariæ).

and on our own coasts cattle eagerly browse on that and kindred
species at low water. In some northern countries, Fucus serratus
sprinkled with meal is used as winter fodder.

All the marine Algæ contain iodine; and even before the value
of that substance in glandular complaints had been ascertained,
stems of a sea-weed were chewed as a remedy by the inhabitants
of certain districts of South America where goître is prevalent.
Chondrus crispus and (Gigartina) mamillosa constitute the Irish
moss of commerce, which dissolves into a nutritious and delicate

jelly, and the restorative value of which in consumption doubtless depends in some degree on the presence of iodine. The freshwater Algæ not only furnish abundant and nourishing food to the fish and other animals living in ponds and streams, but by their action in the decomposition of carburetted hydrogen and other noxious gases purify the element in which they live, thus becoming important sanitary agents. The value of aquatic plants in the aquarium is well known. A Chinese species of Gigartina is much employed as a glue and varnish ; and also much used in China in the manufacture of lanterns and transparencies, and in that country and Japan for glazing windows. Handles for table knives and forks, tools, and other implements have been made from the thick stems of oarweeds, and fishing lines from Chorda filum. Tripoli powder, extensively used for polishing, consists mainly of the silicious shells of Diatoms. On various parts of our coast, the coarser species of sea-weed, now used as a valuable manure, were formerly extensively burnt for kelp, an impure carbonate of soda. This industry, when carried on upon a large scale, became a fruitful source of income to some of the poorest districts in the kingdom, bringing, in the last decade of last century, nearly £30,000 per annum into Orkney alone. Since the production of soda from rock salt has become general, kelp is now only burnt for the extraction of iodine, this being the easiest way of obtaining that substance.

Although the vegetable structure and mode of reproduction are essentially the same in all Algæ, as regards the former they vary from the simple cell, through cells arranged in threads, to a stem and leaves simulating the vegetation of higher tribes. And although the simpler kinds are obviously formed of threads, most of the more compound may also be resolved into the same structure by maceration in hot water or diluted muriatic acid. In substance some are mere masses of slime or jelly, others are silky to the feel, horny, cartilaginous or leather-like, and even apparently woody. A few species secrete carbonate of lime from the water, laying it up in their tissues ; others cover themselves completely with that mineral, while some coat themselves with silex or flint. Many Algæ are beautifully coloured, even when growing at depths to which very little light penetrates. As in their vegetative organs, so in their reproductive, Algæ exhibit many modifications of structure without much real difference. In the green sea-weeds reproduction is effected by simple cell division in the unicellular species, and by spores resulting from the union of the contents of two cells in the others. The red sea-weeds have a double system of reproduction, a distinctly sexual one, by spores and antheridia, and another by tetraspores, which by some are considered to be of the nature of gemmæ, or buds. The spores are generally situated in distinct hollow conceptacles (favellæ, ceramidium, coccidium). The tetraspore is also sometimes contained in a conceptacle. It consists of a more or less globular, transparent cell, which when mature contains within it four (rarely three) sporules. Reproduction in the olive sea-weeds is also double, by zoospores, generally considered gemmæ, and by spores and antherozoids, which is a sexual process.

Following the classification adopted by Professor Harvey, which is that generally employed in English systematic manuals, we divide the order into three sub-orders, named from the pre-

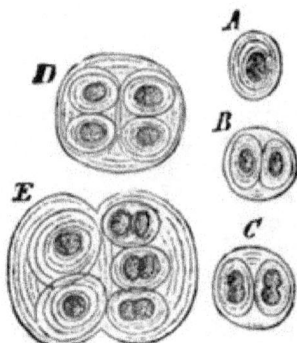

Fig. 2. A, Species of Gleocapsa, one of the Palmelleæ, in various stages. A becomes B, C, D, and E by repeated division. Magnified 300 diameters.

vailing colour of their spores. 1. Chlorospermeæ, with green spores; 2. Rhodospermeæ, with red spores; and 3. Melanospermeæ, with olive-coloured spores. The entire plant in the first group is usually grass-green, but occasionally olive, purple, blue, and sometimes almost black; in the second it is some shade or other of red, very seldom green; and in the third, while generally olive green, it is occasionally brown olive or yellow.

The Chlorospermeæ are extremely varied in form, often threadlike, and are propagated either by the simple division of the contents of their cells (endochrome), by the transformation of particular joints, or by the change of the contents of the cells into zoospores, which are cells moving freely in water by means of hairlike appendages. In their lower forms they are among the most rudimentary of all plants, and thus ot special interest physiologically, as representing the component parts of which higher plants are formed. They are subdivided into twelve groups, as follows :

The first group, Palmelleæ, are unicellular plants, the cells of which are either free or surrounded by a gelatinous mass, and they are propagated by the division of the endochrome. One of the most remarkable of the species of this family is Protococcus cruentus, which is found at the foot of walls having a northern aspect, looking as if blood had been poured out on the ground or on stones. Protococcus nivalis, again, is the cause of the red snow, of which early arctic navigators used to give such marvellous accounts. (Fig. 2.)

The Desmideaceæ, together with the plants of the next succeed-

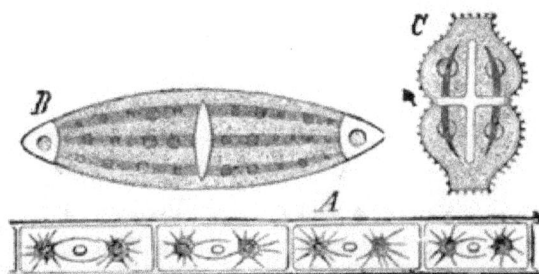

Fig. 3. A, Fragment of a Filament of Zygnema, one of the Conjugateæ. B, Closterium ; C, Euastrium ; two desmids.

ing group, are favourite subjects of investigation or observation by the possessors of microscopes, an attention they merit from the beauty and variety of their forms. They are minute plants of a green colour, consisting of cells generally independent of each other, but sometimes forming brittle threads or minute fronds, and are reproduced by spores generated by the conjugation of two distinct individuals. The process of conjugation in Desmids and Diatoms consists in the union of the endochrome of two individuals, each of which in these families is composed of a single cell. This ultimately forms a rounded body or resting spore, which afterwards germinates, the resulting plant not however acquiring the normal form until the third generation. (Fig. 3.)

The Diatomaceæ, closely allied to the preceding group in structure and reproduction, are however distinguished from them by their flinty shells, which are often beautifully sculptured. Their endochrome is a golden brown, instead of green as in the Desmideaceæ. The latter, also, are confined to fresh water, while the Diatomaceæ are found, though not exclusively, in the sea, where their shells sometimes, microscopically minute as they are individually, form banks extending several hundred miles. It is stated that in the collection made by Sir Joseph Hooker in the Himalayas the species closely resemble our own.

In the next group, Confervaceæ, we are introduced to forms more like the general notion of what a plant should be. The individuals of which it consists are composed of threads, jointed, either simple or branched, mostly of a grass-green colour, and propagating either by minute zoospores or by metamorphosed joints. They are found both in fresh and salt water, and in damp situations. The number of species is very great. A considerable number consist of unbranched threads; the branched forms grow sometimes so densely as to assume the form of solid balls. After floods, when the water stands for several days, they sometimes increase to such an extent, as to form on its subsidence a uniform paper-like stratum, which while decomposing is extremely disagreeable. The name Conferva has been almost discontinued as a generic title, the majority of British species being now ranged under Clado- and Chæto-phora. The latter are branched, and require great care and attention in order to distinguish them, on account of their general resemblance to each other. Good characters are however to be found in their mode of branching and the form and comparative size of the terminal joints.

The Batrachospermeæ constitute a small but very beautiful group, consisting of gelatinous threads variously woven into a branched cylindrical frond. The branches are sometimes arranged, as in the British species, so that the plants appear like necklaces. In colour they pass from green, through intermediate shades of olive and purple, to black. In common with some of the higher Algæ, the threads of the superficial branches send joints down the stem, changing it from simple to compound. The native species are all fluviatile.

The Hydrodicteæ are among the most remarkable of Algæ. Hydrodictyon utriculatum, the solitary British species, is found in the large pond at Hampton Court, and in similar situations in various parts of the country, but not very generally. It resembles a green purse or net, from four to six inches in length, with delicate and regular meshes, the reticulations being about four lines long. Its method of reproduction is no less remarkable than its form. Each of the cells forms within itself an enormous mass of small elliptic grains. These become attached by the extremities so as to form a network inside the cell, and, its walls being dissolved, a new plant is set free to grow to the size of the parent Hydrodictyon.

The Nostochineæ grow in fresh water, or attached to moist soil. They consist of slender, beaded threads surrounded by a firm jelly, and often spreading into large, wavy fronds. The larger beads on the inclosed threads are reproductive spores. (Fig. 4, A.)

The Oscillatoreæ are another remarkable group, on account of the peculiar animal-like motions they exhibit. They occur both in salt and fresh water, and on almost every kind of site in which there is sufficient moisture. The threads of which they are composed are jointed, and generally unbranched ; they are of various tints of blue, red, and green, and, where their fructification has been ascertained, are propagated by cell division.

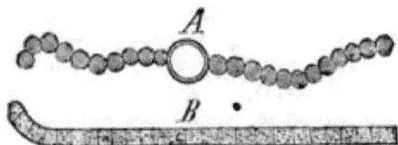

Fig. 4. A, Fragment of a Filament of Nostoc. B, End of a Filament of Oscillatoria.

The most curious point about them is, however, the movements of their fronds. According to Dr. Harvey, these are of three kinds — a pendulum-like movement from side to side, performed by one end, whilst the other remains fixed, so as to form a pivot ; a movement of flexure of the filament itself, the oscillating extremity bending over from one side to the other, like the head of a worm or caterpillar seeking something on its line of march ; and lastly, a simple onward movement of progression, the whole phenomenon being, Dr. Harvey thinks, resolvable into a spiral onward movement of the filament. Whatever is the cause of this motion, it is not, as used to be supposed, of an animal nature ; for the individuals of this group are undoubted plants. (Fig. 4, B.) Several species of Rivularia, belonging to the Oscillatoreæ, are found both in the sea and in fresh water. They are gelatinous, and have something of the appearance of Nostoc, in aspect as well as in minute structure.

The Conjugatæ are freshwater articulated Algæ, which reproduce themselves by the union of two endochromes. They are very interesting objects under the microscope, owing to the spiral or zigzag arrangement of the endochrome of many of them, and the delicacy of all.

The Bulbochæteæ constitute a small group, some half-a-dozen species being British. They are freshwater plants, composed of articulate branched filaments, with fertile bulbshaped branchlets. The endochrome is believed to be fertilized by bodies developed in antheridia, the contents of each fertilized cell dividing into four ovate zoospores.

The last two groups of green sea-weeds consist chiefly of marine plants. Of these the first, Siphoneæ, is so called because the plant, however complicated, is composed invariably of a single cell. It propagates by minute zoospores, by large quiescent spores, or by large active spores clothed with cilia. It includes the remarkable genus Codium, three species of which inhabit the British seas. In Codium Bursa the filamentous frond is spherical and hollow, presenting more the appearance of a round sponge or puff-ball than a sea-weed, and is somewhat rare. Another species greatly resembles a branched sponge, and the third forms a velvety crust on the surface of rocks. Another genus, Vaucheria, is of a beautiful green colour, forming a velvety surface on moist soil, on mud-covered rocks overflowed by the tide, or parasitic on other sea-weeds. The most attractive plants of this family are however those of the genus Bryopsis, two of which are found on the British shores. The most common one is B. plumosa, the fronds of which grow usually in the shady and sheltered sides of rock pools.

The fronds of the last of the green-weed groups, the Ulvaceæ, are membranous, and either flat or tubular. Two of them, Ulva latissima, the green, and Porphyra laciniata, the purple laver, are among the most common sea-weeds, growing well up from low-water mark. The propagation in all of them is by zoospores. An allied genus, Enteromorpha, is protean in its forms, which have been classed under many species. They may, however, be reduced to half a dozen. Some of them are very slender, so as almost to be mistaken for confervoid plants.

With the Rhodospermeæ we enter a sub-order of Algæ, exclusively marine, the plants in which have always held out great attractions to the collector. In structure they are expanded or filamentous, nearly always rose-coloured or purple in colour. Of the fourteen groups into which they are divided by Harvey, the first is Ceramiaceæ, articulate Algæ, constituting a large proportion of the marine plants of our shores. Of the genus Ceramium,

C. rubrum is the most frequent, and it is found in every latitude, almost from pole to pole. It is very variable in aspect, but can always be recognized by its fruit. C. diaphanum is a very handsome species, growing often in rock pools along with the other. There are about fifteen native species altogether, some of them rare, and all very beautiful, both as displayed on paper and seen under the microscope. Crouania attenuata is a beautiful plant, parasitic upon a Cladostephus or Corallina officinalis. It is however extremely rare, being only found in England about Land's

Fig. 5. Species of Callithamnion.

End. A more common and conspicuous, but equally handsome plant is Ptilota plumosa (Fig. 9), which is mostly confined to our northern coasts; although P. sericea, a smaller species, or variety, is common in the south, and easily distinguished from its congener, which it otherwise greatly resembles, by its jointed branchlets and pinnules. Callithamnion, Halurus and Griffithsia, articulate like Ceramium, furnish also several handsome species. (Fig. 5.)

The group Spyridieæ contains only one English plant, Spyridia filamentosa, which is curiously and irregularly branched, the branches being articulate and of a pinky red. One of its kinds of fruit, consisting of crimson spores, is contained in a transparent network basket, formed by the favellæ, or short branches, whence its name.

The Cryptonemiaceæ are very numerous in genera and species. They all have inarticulate branches, some are thread-like. Grateloupia filicina is a neat little plant, met with rarely on the south and west coasts. Gigartina mamillosa, a common plant everywhere, is the plant sold, along with Chondrus crispus, as

Fig. 6. Chondrus crispus.

Irish or Carrageen moss. A handsome little plant, Stenogramme interrupta, is very rare, but it has been gathered both on the Irish and English coasts. The Phyllophoræ, one species of which is frequent on all our shores, may be recognised by the way in which the points and surfaces of their fronds throw out proliferous leaves. Gymnogongrus has two British species, one much resembling Chondrus crispus, already named, of which it was formerly considered a congener. Their fructification is however very different. Ahnfeltia plicata is a curious wiry, entangled plant, almost black in colour, and like horse-hair when dry, and can scarcely be mistaken. Cystoclonium purpurascens is very commonly cast up by the tide on most

Fig. 7. Rhodomenia palmata.

of our coasts. It varies in colour, but is easily distinguished by the spore-bearing tubercles imbedded in its slender branches. Callophyllis laciniata is a handsome species, of a rich crimson colour, and sometimes a foot square. It can scarcely have escaped the notice of the sea-side visitor, for it is widely distributed and often thrown out in great abundance; one writer describes the shore near Tynemouth as having been red for upwards of a mile with this superb sea-weed. Kalymenia reniformis is another of the broad, flat Algæ, but it is scarcer, and of a colour not so conspicuous. Among the most frequent of our sea-weeds, both as growing in the rock pools and cast ashore, is Chondrus crispus, already twice referred to in connexion with its officinal uses. It is very variable in form, one author figuring as many as thirty-six different varieties. (Fig. 6.) Chylocladia clavellosa, which is sometimes cast ashore a fond

Fig. 8. Wormskioldia sanguinea

and a half long, is closely set with branches, and these again clothed with branchlets in one or two series. The whole plant is fleshy, of a rose-red or brilliant pink colour, turning to golden yellow in decay. There is another small species, confined to the extreme north of Britain. Halymenia ligulata is another flat red weed, but sometimes very narrow in its ramifications. Furcellaria fastigiata has a round, branched, taper stem, swollen at the summit, which contains the fruit, consisting of masses of tetraspores in a pod-like receptacle. Schizymenia edulis, better known perhaps by its old name Iridea, is a flat, inversely egg-shaped leaf with scarcely any stem. It is one of the edible

Algæ, and pretty frequent in shady rock pools. Gloiosiphonia capillaris is a remarkably beautiful plant, and not common, being confined to certain parts of the southern coasts. The stem is very soft and gelatinous; the spores are produced in red globular masses imbedded in the marginal filaments, which have a fine appearance under the microscope when fresh.

The Rhodomeniaceæ are purplish or blood-red sea-weeds, inarticulate, membranaceous, and cellular. Among the dark-coloured is Rhodomenia palmata, better known as dulse, a common and edible species. (Fig. 7.) Wormskioldia sanguinea is not only the most beautiful sea-weed, but the finest of all

Fig. 9. Ptilota plumosa.

leaves or fronds. It is usually about six inches long, but sometimes nearly double that length and six inches broad, with a distinct midrib and branching veins, and a delicate wavy lamina, pink or deep red. The fruit is produced in winter from small leaflets growing upon the bare midrib. (Fig. 8.) The commonest of all red seaweeds on our coast, one of the most elegant, and much sought after by sea-weed picture makers, Plocamium coccineum, belongs to this group. Calliblepharis ciliata and jubata are coarser plants, the latter being the more frequent. They were formerly included in the genus Rhodymenia, from which they were removed when their fruit was better understood.

Wrangelia and Naccaria are the only British genera in

Wrangeliaceæ. There is only one native species in each, both being rare, the latter especially.

The Helminthocladiæ are also a limited group, of a gelatinous structure; so much so that on being gathered they feel like a bunch of slimy worms, whence the name of the family. Helminthora purpurea and divaricata with Nemaleon multifidum and Scinaia furcellata represent them in Britain. They are nearly all very rare, pretty plants, and very effective as microscopic objects.

The Squamariæ, formerly included in the Corallinaceæ, are a small group of inconspicuous plants resembling lichens, of a leathery texture, and growing on rocks and shells attached by their lower surface.

A single genus only, Polyides, represents the Spongiocarpeæ. Polyides rotundus resembles Furcellaria fastigiata very closely, but differs widely in the fruit, which consists of spongy warts surrounding the frond, composed of spores and articulated threads.

Of the next group represented in Britain, Gelidiaceæ, we have only one plant, Gelidium corneum, very common on our shores, and perhaps the most variable of all vegetable species.

The Sphærococcidæ include both membranaceous and cartilaginous species. Of the latter is Sphærococcus coronopifolius, which cannot easily be mistaken, owing to the numerous berry-like fruits that tip its branchlets. It is rather rare on the northern, but often thrown ashore in large quantities on the southern coasts. The genus Delesseria has four British species, the largest being the well-known D. sinuosa, the fronds of which resemble an oak leaf in outline. The handsomest are D. ruscifolia and D. hypoglossum, which are more delicate and of a finer colour than sinuosa. There are three British species of Gracillaria, in two of which the branches are cylindrical, and in the other flat. G. compressa makes an excellent preserve and pickle, but unfortunately it is the rarest of the three. Nitophyllum is one of the greatest ornaments of this tribe. There are six British species, which are amongst the most delicate and beautiful of our native Algæ.

The Corallinaceæ are remarkable for the property they possess of absorbing carbonate of lime into their tissues, so that they appear as a succession of chalky articulations or incrustations. The most common is Corallina officinalis. There are two British species of Corallina, and two also of the nearly allied genus, Jania. Of the foliaceous group there are likewise two British genera, Melobesia and Hildenbrantia.

The next group, the Laurenciaceæ, are cartilaginous and cylindrical or compressed, the frond in the greater portion of them being inarticulate and solid. They contain several species valued by collectors, although some of them are amongst our commonest plants. Their colour is, when perfect, a dull purple or brownish red, but they change under the influence of light and air, while fresh water is rapidly destructive to their tints. (Fig. 10.)

Fig. 10. Laurencia pinnatifida.

The Chylocladiæ are curiously jointed plants, removed by Agardh to a new genus, Lomentaria, and a new order Chondriæ. Bonnemaisonia asparagoides is the most rare and beautiful of the tribe.

The last tribe of red weeds, Rhodomelaceæ, varies greatly in the structure of the frond, but the fruit is more uniform. Poly-

B

siphonia and Dasya contain the finest of the filiform division ; the leafy one, Odonthalia, a northern form, is a very beautiful sea-weed both as respects form and colour. Well-grown specimens are not unlike a hawthorn twig, and of a blood red colour.

The plants of the sub-order Melanospermeæ, are, like the red sea-weeds, exclusively marine. They are usually large and coarse, and confined mostly to comparatively shallow water. In the Laminariaceæ we find the gigantic oarweeds already briefly referred to. Lessonia, which encircles in submarine forests the antarctic coasts, is an erect, tree-like plant, with a trunk from five to ten feet high, forked branches, and drooping leaves, one to three feet in length, and has been compared to a weeping willow. Sir Joseph Hooker says, that from a boat there may on a calm day be witnessed in the antarctic regions, over these submarine groves, "as busy a scene as is presented by the coral reefs of the tropics. The leaves of the Lessoniæ are crowded with Sertulariæ and Mollusca, or encircled with Flustra ; on the trunks parasitic Algæ abound, together with chitons, limpets, and other shells ; at the base and among the tangled roots swarm thousands of Crustaceæ and Radiata, while fish of several species dart among the leaves and branches." Of these and other gigantic melanosperms, flung ashore by the waves, a belt of decaying vegetable matter is formed, miles in extent, some yards broad, and three feet in depth ; and Sir J. Hooker adds that the trunks of Lessonia so much resemble driftwood that no persuasion could prevent an ignorant shipmaster from employing his crew, during two bitterly cold days, in collecting this incombustible material for fuel. Macrocystis and Nereocystis are also giant members of this sub-order. Some of the Laminariæ which form a belt around our own coasts not seldom attain a length of from eight to twelve feet. The common bladder-wrack (Fucus vesiculosus) sometimes grows in Jutland to a height of ten feet, and in clusters several feet in diameter. The colour of most of the plants in this sub-order is some shade of olive, but several of them turn to green in drying.

The first group, Ectocarpeæ, is composed of thread-like jointed plants, the fructification of which consists of external spores, sometimes formed by the swelling of a branchlet. The typical genus, Ectocarpus, abounds in species, a dozen or so of which, very nearly allied plants, being found around our own shores. One or two of them are very handsome. There are also some very beautiful plants in the genus Sphacelaria, belonging to this group, several of them resembling miniature ferns. All the Sphacelariæ are easily recognized by the withered appearance of the tips of the fruiting branches. Myriotrichia is a genus of small parasitical plants, the two British species of which grow chiefly on the sea thongs (Chorda).

The Chordariæ are sometimes gelatinous in structure, in other cases cartilaginous. The fruit is contained in the substance of the frond. The genus Chordaria consists of plants which have the appearance of dark coloured twine. There are two British species, one being rather common. Chorda filum, searope, another string-like seaweed, grows in tufts from a few inches to many feet in length, and tapering at the roots to about the thickness of a pig's bristle. In quiet land-locked bays with

Fig. 11. Padina pavonia.

a sandy or muddy bottom, it sometimes extends to forty feet in length, forming extensive meadows, obstructing the passage of boats, and endangering the lives of swimmers entangled in its slimy cords, whence probably its other name of "dead men's lines."

The Mesogloieæ in a fresh state resemble bundles of green, slimy worms. There are three British species, two of which are not uncommon. Although so unattractive in external aspect, they, like many others of the same description, prove very interesting under the microscope. One of the cartilaginous species, Leathsia tuberiformis, has the appearance, when growing, of a mass of distorted tubers.

The species of Elachista, composed of minute parasites, are, as well as unattractive like the Mesogloieæ, inconspicuous, but are beautiful objects when placed under the microscope. Myrionemæ are also parasitic, and even smaller than the plants of the preceding genus.

In the Dictyoteæ the frond is mostly flat, with a reticulated surface, which is sprinkled when in fruit with groups of naked spores or spore cysts. This tribe includes not a few of the most elegant among the Algæ. In structure they are coriaceous, and include plants both with broad and narrow, branched and unbranched fronds. In Haliseris there is a distinct midrib. The largest of the British Dictyoteæ is Cutleria multifida, sometimes found a foot and a half long; and the best known is doubtless Padina pavonia, much sought after by seaside visitors where it grows. Its segments are fan-shaped, variegated with lighter curved lines, and fringed with golden tinted filaments. (Fig. 11.) Owing to its power of decomposing light, its fronds, when growing under water, suggest the train of the peacock, whence its specific name. Taonia atomaria somewhat resembles Cutleria, but exhibits also the wavy lines of Padina. The plant of this group most often cast ashore is Dictyota dichotoma. It makes a handsome specimen when well dried, and is interesting on account of the manner in which it varies in the breadth of its divisions. The variety intricata is curiously curled and entangled. Dictyosiphon fœniculaceus, the solitary British example of its genus, is a bushy filiform plant, remarkable for the beautiful net-like markings of its surface. The Punctariæ have flattened fronds, marked with dots, which sufficiently distinguish them from all the others. A small form is often found parasitic on Chorda filum, spreading out horizontally like the hairs of a bottle brush. Asperococcus derives its name from its roughened surface, occasioned by the thickly scattered spots of fructification.

The Laminariaceæ are inarticulate, mostly flat, often strap-shaped. Their spores occur in superficial patches, or covering the whole frond. The plants of this order, as we have already seen, include the giants of submarine vegetation. In point of mass they constitute the larger part of our native Algæ, although they number only a few species. They are popularly known as tangle or oarweeds, and the stems of Laminaria saccharina and the midrib of Alaria esculenta are used as food.

The Sporochnaceæ are a small but beautiful tribe, inarticulate, and producing their spores in jointed filaments or knob-like masses, and remarkable for their property of turning from olive brown to a verdigris green when exposed to the atmosphere.

They are deep sea plants, or at least grow about low water mark. The largest of the group is Desmarestia ligulata, which, with the other British species, D. aculeata, is often cast ashore. The latter species, at an early period of its existence, is

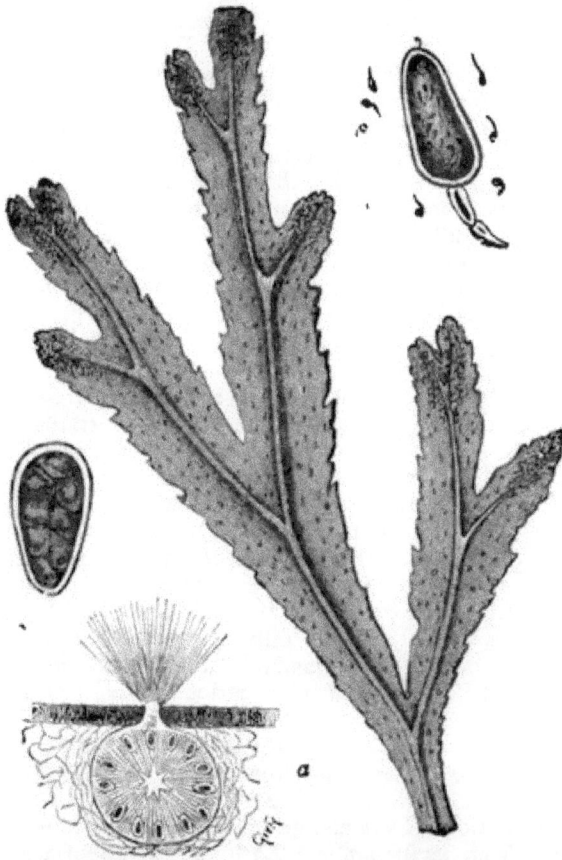

Fig. 12. Fucus serratus, showing a transverse section of the Conceptacle, and Antheridium with Antherozoids escaping.

clothed with tufts of slender hairs, springing from the margin of the frond. Desmarestia viridis is the most delicate and also the rarest of the three. Nothing like fruit has been discovered on any of them. Arthocladia villosa and Sporochnus pedunculatus are branched sea weeds, covered also with tufts of closely set hairs. Carpomitra Cabreræ, a rare species, bears, in common with the two preceding species, its spores in a special receptacle. In the first the receptacle is pod-like ; in the second knotted ; and in the last mitriform.

The concluding group of Algae is the Fucaceæ, including the universally known sea wrack (Fucus). The frond in all of them is jointless. They are reproduced by means of antheridia and oogonia developed in conceptacles, clustered together at the apex of the branches. Both from their bulk and their decided sexual distinctions, they deserve to rank at the head of the order. Of all sea-weeds they are also perhaps of the greatest use to man. One of the most interesting among them is the Gulfweed (Sargassum bacciferum), occupying a tract of the Atlantic extending over many degrees of latitude. Pieces of it, and of its congener, S. vulgare, are occasionally drifted to our shores, and they consequently find a place in works on British Algæ, although they have no claim to be considered native plants. On rocky coasts the various species of Fucus occupy the greater part of the space between tide-marks, the most plentiful being Fucus vesiculosus. F. serratus (Fig. 12) is the handsomest of the genus, the other species being F. nodosus, said to be the most useful for making kelp, and F. canaliculatus. Halidrys siliquosa is remarkable for its spore receptacles, which have quite the appearance of the seed vessel of a flowering plant. The species of Cystoseira, chiefly confined to the southern coasts, are also very interesting. Their submerged fronds are beautifully iridescent, and the stems, of the largest species at least, are generally covered with a great variety of parasites, animal and vegetable, the former consisting of Hydrozoa and Polyzoa, and other curious forms. Himanthalia lorea is another remarkable plant. It has conspicuous forked fruit-bearing receptacles ; but the real plants are the small cones at the base of these, and from which they are shed when ripe.

As to conditions of site and geographical distribution, Algæ do not differ from land plants. Latitude, depth of water, and currents influence them in the same way as latitude, elevation, and station operate on the latter ; and the analogy is maintained in the almost cosmopolitan range of some, and the restricted habitat of others. Not many extra-European species of Desmids are known, but those of Diatoms are far more widely diffused, and extend beyond the limits of all other vegetation, existing wherever there is water sufficient to allow of their production ; and they are found not only in water, but also on the moist surface of the ground and on other plants, in hot springs and amid polar ice. They are said to occur in such countless myriads in the South Polar Sea as to stain the berg and pack ice wherever these are washed by the surge. A deposit of mud, chiefly consisting of the shells of Diatoms, 400 miles long, 120 miles broad, and of unknown thickness, was found at a depth of between 200 and 400 feet on the flanks of Victoria Land in 70° south latitude. Such is their abundance in some rivers and estuaries that Professor Ehrenberg goes the length of affirming that they have exercised an important influence in blocking up harbours and diminishing the depth of channels. The trade and other winds distribute large quantities over the earth, which may account for the universality of their specific distribution ; for Sir Joseph Hooker found the Himalayan species to closely resemble our own. Common British species also occur in Ceylon, Italy, Virginia, and Peru. The typical species of the Confervaceæ are also distributed over the whole surface of the globe. They inhabit both fresh and salt water, and are found alike in the polar seas and in the boiling springs of Iceland, in mineral waters and in chemical solutions. Some of the tropical ones are exceedingly large and dense. Batrachospermum vagum, in the next tribe, a native of England, is also found in New Zealand. An edible species of Nostochineæ, produced on the boggy slopes bordering the Arctic Ocean, is blown about by the winds sometimes ten miles from land, where it is found lying in small depressions in the snow upon the ice. The common Nostoc of moist ground in England occurs also in Kerguelen's Land, high in the southern hemisphere. Floating masses of Monormia are often the cause of the green hue assumed by the water of ponds and lakes. Certain species of Oscillatoria of a deep red colour live in hot springs in India, and the Red Sea is supposed to have derived its name from a species of this tribe, which covers it with a scum for many miles, according to the direction of the wind. The lake of Glaslough in County Monaghan, Ireland, owes its colour and its name to Oscillatoria ærugescens, and large masses of water in Scotland and Switzerland are tinted green or purple by a similar agency. A few species of Siphoneæ have a very wide range, two British species of Codium occurring in New Zealand. The Ulvaceæ abound principally in the colder latitudes. Enteromorpha intestinalis, a common British species, is as frequent in Japan, where it is used, when dried, in soup. The Rhodosperms are found in every sea, although the geographical boundaries of genera are often well-marked. Gloiosiphonia, one of our rarest and most beautiful Algæ, is widely diffused. Of Melanosperms the Laminariæ affect the higher northern latitudes, Sargassa abound in the warmer seas, while Durvillæa, Lessonia, and Macrocystis characterize the marine flora of the Southern Ocean. The Fucaceæ are most abundant towards the poles, where they attain their greatest size. The marine meadows of Sargassum, conceived by some naturalists to mark the site of the lost Atlantis, and which give its name to the Sargasso Sea, extending between 20° and 25° north latitude, in 40° west longitude,

Fig. 13.   The Gulf-weed (Sargassum bacciforum).

occupy now the same position as when the early navigators, with considerable trepidation, forced through their masses on the way to the New World. Sargassum is drifted into this tract of ocean by currents, the plants being all detached ; and they do not produce fruit in that state, being propagated by buds, which originate new branches and leaves. (Fig. 13.)

Owing to their soft, cellular structure, Algæ are not likely to be preserved in a fossil state ; but what have been considered such have been found as low down as the Silurian formation, although their identity has been disputed, and several of them, it is more than probable, belong to other orders, and some even to the animal kingdom. Freshwater forms, all of existing genera and species, are believed to have been detected in the carboniferous rocks of Britain and France ; others also of the green-coloured division are said to occur from the Silurian to the Eocene, and the Florideæ to be represented from the Lias to the Miocene. The indestructible nature of the shells of the Diatomaceæ has enabled them to survive where the less protected species may have perished. Tripoli stone, a Tertiary rock, is entirely composed of the remains of microscopic plants of this tribe. It is from their silicious shells that mineral acquires its use in the arts, as powder for polishing stones and metals. Ehrenberg estimates that in every cubic inch of the tripoli of Bilin, in Bohemia, there are 41,000,000 of Gaillonella distans. Districts recovered from the sea frequently contain myriads of Diatoms, forming strata of considerable thickness ; and similar deposits occur in the ancient sites of lakes in this and other countries.

Before setting out in search of Algæ the collector ought to provide himself with a pair of stout boots to guard his feet from the sharp-pointed rocks, as well as a staff or pole to balance himself in rock-climbing, which ought to have a hook for drawing floating weed ashore. A stout table-knife tied to the other end will be found very useful. A basket—a fishing-basket does very well—or a waterproof bag, for stowing away his plants, is also necessary. It is advisable to carry a few bottles for the very small and delicate plants, and care should be taken to keep apart, and in sea-water, any specimens of the Sporochnaceæ ; for they are not only apt to decay themselves but to become a cause of corruption in the other weeds with which they come in contact. These bottles should always be carried in the bag or pocket, never in the hand.

Sea-weeds, as every visitor to the coast knows, are torn up in great numbers by the waves, especially during storms, and afterwards left on the shore by the retiring tide. Many shallow-growing species are also to be found attached to the rocks, and in the rock pools, between high and low water mark. There are three points on the beach where the greatest accumulations of floating Algæ are found : high water mark, mid tide level, and low water mark. Low water occurs about five or five and a half hours after high water. The best time for the collector to commence is half an hour or so before dead low

water. He can then work to the lowest point safely, and, retiring before the approaching tide, examine the higher part of
the beach up to high water mark. If the coarse weeds in the
rock pools and chinks are turned back, many rare and delicate
Algæ will be found growing under them, especially at the lowest
level. The most effective method of collecting the plants of
deeper water is by dredging, or going round with a boat at the
extreme ebb, and taking them from the rocks and from the
Laminaria stems, on which a great number have their station.
Stems of Laminaria thrown out by the waves should also be
carefully examined. In all cases the weed should be well rinsed
in a clear rock pool before being put away in the bag or other
receptacle.

The next thing to be considered is the laying out and preserving of the specimens selected for the herbarium. Wherever
possible these should be laid out on paper, and put under pressure as soon as gathered, or on the same day at all events.
When this is impracticable, they may be spread between the
folds of soft and thick towels and rolled up. Thus treated the
most delicate plants will keep fresh until next day. Another
way is to pack the plants in layers of salt, like herrings ; but
the most usual method of roughly preserving sea-weeds collected
during an unprepared visit to the shore is by moderately drying
them in an airy room out of the direct rays of the sun. They
are then to be placed lightly in bags, and afterwards relaxed by
immersion and prepared in the usual way. The finer plants,
however, suffer more or less by this delay. If carried directly
home from the sea the plants should be emptied into a vessel of
sea-water. A flat dish, about fourteen inches square and three
deep, is then to be filled with clean water. For most plants
this may be fresh, for some it is essential that it should be salt.
Some of the Polysiphonias and others begin to decompose at
once if placed in fresh water. The Griffithsias burst and let
out their colouring matter, and a good many change their colour.
The appliances required are some fine white paper—good printing demy, thirty-six pounds or so in weight per ream, does very
well,—an ample supply of smooth blotting paper, the coarse
paper used by grocers and called "sugar royal," or, best of
all, Bentall's botanical drying paper, pieces of well-washed
book muslin, a camel's hair brush, a bodkin for assisting
to spread out the plants, a pair of scissors, and a pair of
forceps. The mounting paper may be cut in three sizes :
5 in. by 4 in., 7½ in. by 5¼ in., and 10 in. by 7½ in. Then
having selected a specimen, place it in the flat dish referred

to above, and prune it it necessary. Next take a piece of the mounting paper of suitable size, and slip it into the water underneath the plant, keeping hold of it with the thumb of the left hand. Having arranged the plant in a natural manner on the paper, brush it gently with the camel's hair pencil, to remove any dirt or fragments, draw out paper and plant gently and carefully in an oblique direction, and set them on end for a short time to drain. Having in this way transferred as many specimens as will cover a sheet of drying paper, lay them upon it neatly side by side, and cover them with a piece of old muslin. Four sheets of drying paper are then to be placed upon this, then another layer of plants and muslin and four more sheets of drying paper, until a heap, it may be six or eight inches thick, is built up. Place this between two flat boards, weighted with stones, bricks, or other weights ; but the pressure should be moderate at first, otherwise the texture of the muslin may be stamped on both paper and plant. The papers must be changed in about three hours' time, and afterwards every twelve hours. In three or four days, according to the state of the weather, the muslin may be removed, the plants again transferred to dry paper, and subjected to rather severe pressure for several days.

The very gelatinous plants require particular treatment. One way is to put them in drying paper and under a board but to apply no other pressure, change the drying paper at least twice during the first half hour, and after the second change of dryers apply very gentle pressure, increasing it until the specimens are fully dry. A safer and less troublesome way, for the effi- cacy of which we can vouch, is to lay down the plants and dry them without any pressure, afterwards damping the back of the mounting papers and placing them in the drying press. Some Algæ will scarcely adhere to paper. These should be pressed until tolerably dry, then be immersed in skim-milk for a quarter of an hour, and pressed and dried as before. A slight applica- tion of isinglass, dissolved in alcohol, to the under side of the specimen is sometimes necessary. Before mounting, or at all events before transference to the herbarium, care should be taken to write in pencil on the back of the paper the name of the plant, if known, the place where gathered, and the date. The coarse olive weeds, such as the bladder-wrack, Halidrys, and the like, may in the case of a short visit to the coast be allowed to dry in an airy place, and taken home in the rough. Before pressing, in any case, they should be steeped in boiling water for about half an hour to extract the salt, then washed in clean fresh water, dried between coarse towels, and pressed and dried in the same way as flowering plants. A collection of Algæ may be fastened on sheets of paper of the usual herbarium size and kept in a cabinet or portfolios, or attached to the leaves of an album. For scientific purposes, however, the latter is the least convenient way.

There are few objects more beautiful than many of the sea-weeds when well preserved; but the filiform species, especially those of the first sub-order, do not retain their distinguishing characters when pressed as has been described. Portions of these, however, as well as sections of stems and fruit, may be usefully dried on small squares of thin mica, for subsequent microscopic examination, or they may be mounted on the ordinary microscope slides. This is the only course possible with Desmids and Diatoms. The former are to be sought in shallow pools, especially in open boggy moors. The larger species commonly lie in a thin gelatinous stratum at the bottom of the pools, and by gently passing the fingers under them they will be caused to rise towards the surface, when they can be lifted with a scoop. Other species form a greenish or dirty cloud on the stems and leaves of other aquatic plants, and by stripping the plant between the fingers these also may be similarly detached and secured. If they are much diffused through the water, they may be separated by straining through linen; and this is a very common way of procuring them. Living Diatoms are found on aquatic plants, on rocks and stones, under water or on mud, presenting themselves as coloured fringes, cushion-like tufts, or filmy strata. In colour the masses vary from a yellowish brown to almost black. They are difficult, both when living and dead, to separate from foreign matter; but repeated washings are effectual in both cases, and, for the living ones, their tendency to move towards the light may also be taken advantage of. When only the shells are wanted for mounting, the cell contents are removed by means of hydrochloric and nitric acid. The most satisfactory medium for preserving fresh Desmids and Diatoms is distilled water, and if the water is saturated with camphor, or has dissolved in it a grain of alum and a grain of bay salt to an ounce of water, confervoid growths will be prevented. For larger preparations of Algæ, Thwaites' fluid is strongly recommended. This is made by adding to one part of rectified spirit as many drops of creasote as will saturate it, and then gradually mixing with it in a pestle and mortar some prepared chalk, with sixteen parts of water; an equal quantity of water saturated with camphor is then to be added, and the mixture, after standing for a few days, to be carefully filtered.

For authorities on the morphology and classification of the Algæ, students may be referred to Sachs' "Text Book" and Le Maout's "System of Botany," of which there are good translations, and the "Introduction to Cryptogamic Botany," by the Rev. M. J. Berkeley; for descriptions and the identification of species, to the text and figures of Harvey's "Phycologia Britannica," and "Nature-Printed Sea-weeds." Both of these are however costly. Among the cheaper works are "British Sea-weeds," by S. O. Gray (Lovell, Reeve & Co.), "Harvey's Manual" and an abridgment by Mrs. A. Gatty, with reduced but well executed copies of the figures, of the Phycologia. This synopsis can often be picked up cheap at second-hand bookstalls ; and there is a very excellent low-priced work suitable for amateurs, Grattann's "British Marine Algæ," containing recognizable figures of nearly all our native species. Landsborough's "Popular History of British Sea-weeds," and Mrs. Lane Clarke's "Common Sea-weeds," are also cheap and useful manuals on the subject.

SHELLS.

BY

B. B. WOODWARD.

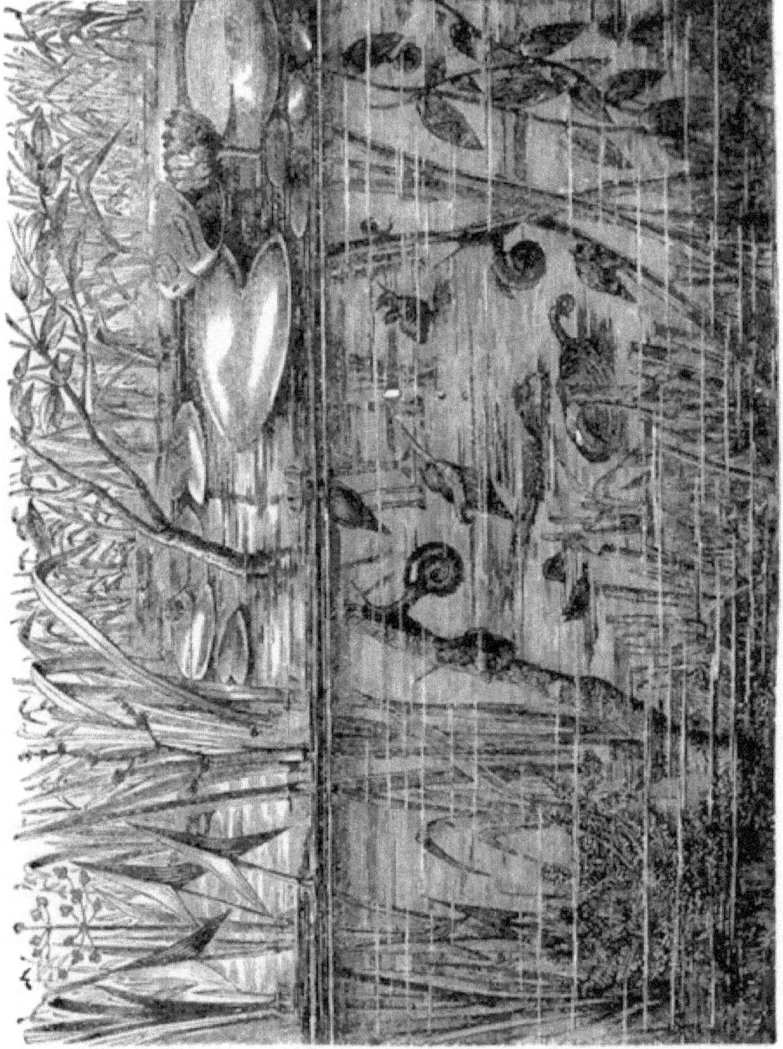

POND SNAILS.

SHELLS.

INTRODUCTORY.

IN the very earliest times, long before there was any attempt
at the scientific classification and arrangement of shells, they
appear to have been objects of admiration, and to have been
valued on account of their beauty, for we find that the pre-
historic men, who, in company with the mammoth, or hairy
elephant, and other animals now extinct, inhabited Southern
France in days long gone by, used to bore holes in them, and,
like the savage of to-day, wear them as ornaments. The Greek
physician and philosopher, Aristotle, is said to have been the
first to study the formation of shells, and to raise the knowledge
thus acquired into the position of a science ; by him shells were
divided into three orders—an arrangement preserved, with some
small changes, by Linnæus.   It is possible that the world-wide
renown of the Swedish naturalist during the last century, and
the ardour with which he pursued his investigations, may have
given an impetus to the study of natural objects, for we find that
at that period large sums were often given by collectors for
choice specimens of shells.   Nor is this to be wondered at, for
few things look nicer, or better repay trouble expended on them,
than does a well-arranged and carefully mounted and named
collection of shells.   Certainly nothing looks worse than a
number of shells of all descriptions, of every kind, shape, and
colour, thrown promiscuously into a box, like the unfortunate

animals in a toy Noah's ark, to the great detriment of their value and beauty; for, as the inevitable result of shaking against each other, the natural polish is taken off some, the delicate points and ornaments are broken off others, the whole collection becoming in time unsightly and disappointing, and all for want of a little care at the outset.

In this, as in every other undertaking, "how to set about it" is the chief difficulty with beginners; and here, perhaps, a few hints gathered from experience may not be without value. One thing a young collector should always bear in mind, however, is, that no instructions can be of any avail to him unless, for his part, he is prepared to bring patience, neatness, and attention to detail, to bear upon his work.

Since it is important to know the best way of storing specimens already acquired, we will, in the first place, devote a few words to this point, and then proceed to describe the best means of collecting specimens, and of naming, mounting, and arranging the same.

HOW TO MAKE A CABINET.

It is a common mistake, both with old and young, to imagine that a handsome cabinet is, in the first instance, a necessity; but no greater blunder can be made: the cabinet should be considered merely an accessory, the collection itself being just as valuable, and generally more useful, when kept in a series of plain wooden or cardboard boxes. We intend, therefore, to describe the simplest possible means of keeping a collection of shells, leaving elaborate and costly methods to those who value the case more than its contents.

The first thing required is some method of keeping the different species of shells apart, so that they may not get mixed, or be difficult to find when wanted. The simplest plan of doing this is to collect all the empty chip match-boxes you can find, throw away the cases in which they slide, and keep the trays, trying to get as many of a size as possible. (The ordinary Bryant & May's, or Bell & Black's, are the most useful, and with them the trays of the small Swedish match-boxes, two of which, placed side by side, occupy nearly exactly the same space as one and a half of the larger size, and so fit in with them nicely.) In these trays your shells should be placed, one kind in each tray; but although very convenient for most specimens, they will of course be too small for very many, and so the larger trays must be made. This may easily be done as follows: cut a rectangular piece of cardboard two inches longer one way than the length of the match-tray, and two inches more the other way than twice the width of the match-tray; then with a pencil rule lines one inch from the edges and parallel with them (Fig. 1); next cut out the little squares (*a a*, *a a*) these lines form in the corners of the piece of cardboard, and then with a penknife cut *half* through the card, exactly on the remaining pencil-lines, and bend up the pieces, which will then form sides for your tray; and by binding it round with a piece of blue paper, you will have one that will look neat, uniform with the others, and yet be just twice their size. If required, you can make in the same way any size, only take care that they are all multiples of one standard size, as loss of space will thereby be avoided when you come to the next process in your cabinet. This is, to get a large box or tray in which to hold your smaller ones.

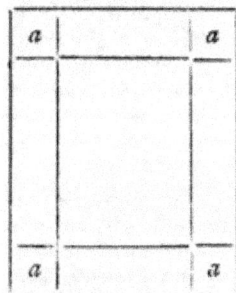

Fig. 1. How to cut a cardboard tray.

The simplest plan is to get some half-dozen cardboard boxes (such as may be obtained for the asking or for a very trifling cost at any draper's), having a depth of from one to two inches (according to the size of your shells); in these your trays may be arranged in columns, and the boxes can be kept one above the other in a cupboard or in a larger box. More boxes and trays can, from time to time, be added as occasion requires, and thus the whole collection may be kept in good working order at a trifling cost. A more durable form of cheap cabinet may be made by collecting the wooden boxes so common in grocers' shops, cleaning them with sand-paper, staining and varnishing them outside, and lining them inside with paper; or, if handy at carpentering, you may make all your boxes, or even a real cabinet, for yourself.

HOW TO COLLECT SHELLS.

Provision being thus made for the comfortable accommodation of your treasures, the next consideration is, how to set about collecting them. Mollusca are to be found all over the globe, from the frozen north to the sun-baked tropics, on the land or in lakes, rivers, or seas—wherever, in fact, they can find the food and other conditions suitable for their growth and development ; but the collector who is not also a great traveller, must of course rely for his foreign specimens upon the generosity of friends, or else procure them from dealers. In most districts of our own country, there are, however, to be found large numbers of shells whose variety and beauty will astonish and reward the efforts of any patient seeker. Begin with your own garden,—search in the out-of-the-way, and especially damp, corners ; turn over the flower-pots and stones which have lain longest in one place, search amongst the roots of the grass growing under walls, and in the moss round the roots of the trees, and you will be sur-prised at the number of different shells you may find in a very short space of time. When the resources of the garden have been exhausted, go into the nearest lanes and again search the grass and at the roots of plants, especially the nettles which grow beside ditches and in damp places ; hunt amongst the dead leaves in plantations, and literally leave no stone unturned. All the apparatus it is necessary to take on these excursions consists of a few small match or pill-boxes in which to carry home the specimens ; a pair of forceps to pick up the smaller ones, or to get them out of cracks ; a hooked stick to beat down and pull away the nettles ; and, above all, sharp eyes trained to powers of observation. The best time to go out, is just after a warm shower, when all the grass and leaves are still wet, for the land-snails are very fond of moisture, and the shower entices them out of their lurking-places. Where the ground is made of chalk or limestone, they will be found most abundant ; for as the snail's shell is composed of layers of animal tissue, strengthened by depositions of calcareous earthy-matter which the creature gets from the plants on which it feeds, and these in their turn obtain from the soil—it naturally follows that the snail prefers to dwell where that article is most abundant, as an hour's hunt on any chalk-down will soon show.

When garden and lanes are both exhausted, you may then turn to the ponds and streams in the neighbourhood, where you will find several new kinds. Some will be crawling up the rushes near the margin of the water, others will be found in the water near the bank, while others may be obtained by pulling on shore pieces of wood and branches that may be floating in the water ; but the best are sure to be beyond the reach of arm or stick, and it will be necessary to employ a net, which may be easily made by bending a piece of wire into a circle of about four inches in diameter, and sewing to it a small gauze bag ; it may be mounted either on a long bamboo, or, better still, on one of those ingenious Japanese walking-stick fishing-rods. For heavier work, however, such as getting fresh-water mussels and other mollusca from the bottom, you will require a net something like the accom-

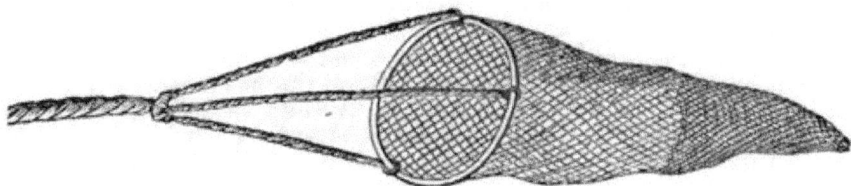

Fig. 2. Net for taking water-snails.

panying figure (Fig. 2), about one foot in diameter. This, when attached to a long rope, may be thrown out some distance and dragged through the water-weeds to the shore, or if made with a square instead of a circular mouth, it may be so weighted that it will sink to the bottom, and be used as a dredge for catching the mussels which live half-buried in the mud. To carry the water-snails home, you will find it necessary to have tin boxes (empty mustard-tins are the best), as match-boxes come to pieces when wetted.

The finest collections of shells, however, are to be made at the sea-side, for the marine mollusca are both more varied in kind and more abundant than the land and fresh-water ones, and quite an extensive collection may be made in the course of an afternoon's ramble along the shore; it is necessary, however, to carefully reject such specimens as are worn by having been rolled by the waves upon the beach, as they are not of any great value in a collection; it is better, in fact, if possible, to go down to the rocks at low water and collect the living specimens. Search well about and under the seaweeds, and in the rock-pools, and, when boating, throw your dredge-net out and tow it behind, hauling it in occasionally to see what you have caught, and to empty the stones and rubbish out.

At low tide also, look out for rocks with a number of round holes in them, all close together, for in these holes the Pholas (Fig. 22) dwells, having bored a burrow in the solid rock, though *how* he does it we do not yet quite know.

The Razor-shells and Cockles live in the sand, their presence being indicated by a small round hole ; but they bury themselves so fast that you will find it difficult to get at them. Some good specimens, too, of the deeper water forms are sure to be found near the spots where fishermen drag their boats ashore, as they are often thrown away in clearing out the nets ; moreover, if you can make friends with any of the said fishermen, they will be able to find and bring you many nice specimens from time to time.

The reason that so much has been said about collecting living specimens, is not only because in them the shell is more likely to be perfect, but also because in its living state the shell is coated with a layer of animal matter, sometimes thin and transparent, at others thick and opaque, called the *periostracum* (or *epidermis*), which serves to protect the shell from the weather, but which perishes with the animal, so that dead shells which have lain for some time tenantless on the ground, or at the bottom of the water, exposed to the destructive agencies that are constantly at work in nature, have almost invariably lost both their natural polish and their varied hues, and are besides only too often broken as well. Since, however, even a damaged specimen is better than none at all, such should always be kept until a more perfect example can be obtained.

HOW TO PREPARE THE SHELLS FOR THE CABINET.

The question with which we have next to deal is, after collecting a number of living mollusks, how, in the quickest and most painless manner possible, to kill the animals in order to obtain possession of their shells. There is but one way we know of in which this may be accomplished, and that is by placing the creatures in an earthen jar and pouring *boiling* water on them. With land, or fresh-water snails, the addition of a large spoonful of table-salt is advisable, as it acts upon them chemically, and not only puts them sooner out of pain, but also renders their subsequent extraction far easier. Death by this process is instantaneous, and consequently painless ; but to leave snails in cold salt water is to inflict on them the tortures of a lingering death ; while for the brutality of gardeners and other thoughtless persons who seek to destroy the poor snail they find eating their plants by crushing it under foot on the gravel path, no words of condemnation are too strong, since it must always be borne in mind that snails have not, like us, *one* nervous centre, but three, and are far more tenacious of life ; hence, unless all the nerves are destroyed at once, a great deal of suffering is entailed on the poor creature ; and if merely crushed under foot, the mangled portions *will live for hours.* Hot water has also the advantage of tending to remove the dirt which is almost sure to have gathered on the shells, and so helping to prepare them better for the cabinet. As soon as the water is cool enough, fish out the shells one by one and proceed to extract the dead animals. This, if the mollusk is *univalve* (*i.e.*, whose shell is composed of a single piece), such as an ordinary garden snail, can easily be done by picking them out with a pin ; you will find, probably, that some of the smaller ones have shrunk back so far into their shells as to be beyond the reach of a straight pin, so it will be necessary to bend the pin with a pair of pliers, or, if none are at hand, a key will answer the purpose if the pin be put into one of the notches and bent over the edge until sufficiently curved to reach up the shell. You will find it convenient to keep a set of pins bent to different curves, to which you may fit handles by cutting off the heads and sticking them into match stems. It is a good plan to soak some of the smaller snails in clean cold water before killing them, as they swell out with the water, and do not, when dead, retreat so far into their shells. If you have a microscope, and wish to keep the animals till you have time to get the tongues out, drop the bodies into small bottles of methylated spirit and water, when they will keep till required, otherwise they should of course be thrown away at once. The now empty shells should be washed in clean warm water, and, if very dirty, gently scrubbed with a soft nail or tooth brush, and then carefully dried.

In such shells as the Periwinkle, Whelk, etc., whose inhabitants close the entrance of their dwelling with a trap-door, or *operculum* as it is called, you should be careful to preserve each with its proper shell.

If you are cleaning *bivalves*, or shells composed of two pieces, like the common mussel, you will have to remove the animal with a penknife, and while leaving the inside quite clean, be very careful not to break the ligament which serves as a hinge ; then wash as before, and tie them together to prevent their gaping open when dry.

Sometimes the fresh-water or marine shells are so coated over with a vegetable growth that no scrubbing with water alone will remove it, and in these cases a weak solution of caustic soda may be used, but very carefully, since, if too strong a solution be employed, the surface of the shell will be removed with the dirt, and

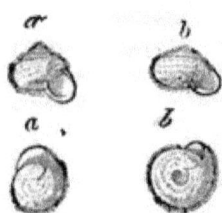

Fig. 3. (a) *Helix sericea* and (b) *Helix hispida.*

the specimen spoilt. In some shells the periostracum is very thick and coarse, and must be removed before the shell itself can be seen; but it is always well to keep at least one specimen in its rough state as an example. In other shells the periostracum is covered over with very fine, delicate hairs (*Helix sericea* and *Helix hispida,* Fig. 3), and great care must then be taken not to brush these off.

HOW TO MOUNT THE SHELLS FOR THE CABINET.

When the specimens are thoroughly cleaned, the next process is to sort out the different kinds, placing each description in a different tray, and then to get them ready for mounting, for no collection will look well unless each kind is so arranged that it may be seen to the best advantage, and is also carefully named. Where you have a good number, pick out first the largest specimens of their kind, then the smallest, then a series, as you have room for them, of the most perfect ; and finally those which show any peculiarity of structure or marking. Try, too, to get young forms as well as adult, for the young are often very different in appearance from the full-grown shell. Mark on them, especially on such as you have found yourself, the locality they came from, as it is very important to the shell collector to know this, since specimens common enough in one district are often rare in another. Either write the name of the place in ink on a corner of the shell itself, or gum a small label just inside it, or simply number it, and write the name of the place with a corresponding number against it in a book kept for the purpose. Next select a tray large enough to hold all you have of this kind ; place a piece of cotton wool at the bottom, and lay your shells upon it. For small shells, however, this method is not suitable, as the cotton wool acts on them like a spring mattress, and they are liable on the least shock to be jerked out of their trays and lost. This difficulty may be met by cutting a piece of cardboard so that it just fits into your tray, and then gumming the shells on to it in rows ; but remember that, in this plan of mounting, it is impossible to take the shells up and examine them on all sides as you do the loose ones, and so you must mount a good many, and place them in many different positions, so that they may be seen from as many points of view as possible. The gum used should always have nearly one-sixth of its bulk of pure glycerine added to it ; this prevents it from becoming brittle when dry, otherwise your specimens would be liable after a time to break away from the card and get lost. If the shells will not stay in the position you require, wedge them up with little pieces of cork until the gum is dry.

When the shells are mounted, you must try, if you have not already done so, to get the proper names for them ; it is as important to be able to call shells by their right names as it is to know people by theirs. The commoner sorts you will be able to name from the figures of them given in text-books, such as those quoted in the list at the end of this little work ; but some you will find it very difficult to name, and it will then be necessary to ask friends who have collections to help you, or to take them to some museum and compare them with the named specimens there exhibited. When the right name is discovered, your label must then be written in a very small, neat hand, and gummed to the edge of the tray or on the card if your specimens are mounted. At the top you put the Latin name, ruling a line underneath it, and then, if you like, add the English name ; next, put the name of the place and the date at which it was found, thus :—

Helix aspersa (Common snail),

Lane near Hampstead Heath,
July 10th, 1882.

A double red ink line ruled at the top and bottom will add a finished appearance to it.

HOW TO CLASSIFY THE SHELLS FOR THE CABINET.

All the foregoing processes, except the naming of your specimens, are more or less mechanical, and are only the means to the end—a properly arranged collection. For, however well a collection may be mounted, it is practically useless if the different shells composing it be not properly classified. By classification is meant the bringing together those kinds that most resemble each other, first of all into large groups having special characteristics in common, and then by subdividing these into other smaller groups, and so on. Thus the animal kingdom is divided, first of all, into *Sub-kingdoms*, then each *Sub-kingdom* into so many *Classes* containing those which have further characteristics in common, the *Classes* into *Orders*, the *Orders* into *Families*, the *Families* into *Genera*, and these again into species or kinds.

The Mollusca, or soft-bodied animals, of whose protecting shells your collection consists, form a sub-kingdom, and are subdivided into four classes :—

1. Cephalopoda.
2. Gasteropoda.
3. Pteropoda.
4. Lamellibranchiata (or Conchifera).

And these again into Families, Genera, and Species.

The space at our disposal being limited, it is impossible to do more than furnish some general outlines of the different forms. For further details it will be necessary to refer to one of the larger works, a list of which will be found on the last page.

CLASS I.—The CEPHALOPODA (Head-footed) contains those mollusca that, like the common Octopus, have a number of feet (or arms) set round the mouth, and is divided into those having two gills. (Order I. Dibranchiata) ; and those with four (Order II. Tetrabranchiata). Order I. is again divided into : (*a.*) Those with *eight* feet like the Argonaut (or Paper-nautilus, Fig. 4), which fable has so long endowed with the power of sailing on the surface of the ocean, (it is even represented in one book as propelling itself through the air !) and the common Octopus. (*b*) Those with *ten* feet, such as the Loligo (or Squid, Fig. 6), whose delicate internal shell so much resembles a pen in shape; the Cuttle-fish (Sepia, Figs. 5 & 7),

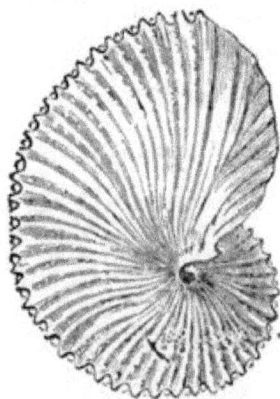

Fig. 4. *Argonauta Argo.*

Fig. 5. "Bone" of *Sepia officinalis.*

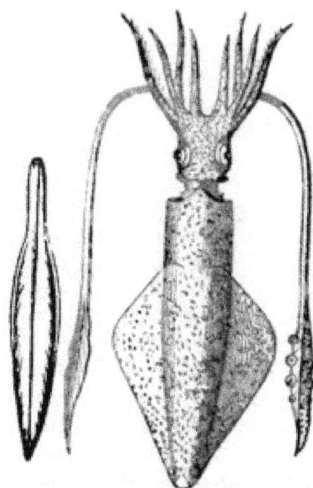

Fig. 6. *Loligo vulgaris*, and " Pen." Fig. 7. *Sepia officinalis.*

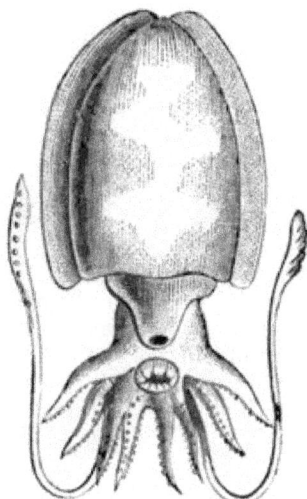

whose so-called " bone " (once largely used as an ink eraser) is frequently found on our southern coasts ; and the pretty little *Spirula* (Fig. 8).

The only representative of the four-gilled order now living is the well-known Pearly Nautilus ; but in former times the Tetrabrarchiata were extremely numerous, especially the *Ammonites.*

CLASS II.—GASTEROPODA (Belly-footed) comprises those mollusca which, like the common snail, creep on the under-surface of the body, and with one exception (*Chiton*, Fig. 20) their shells are univalve (*i.e.*, composed of one piece).   But before we go further, it may be well to point out the names given to different parts of a univalve shell. The aperture whence the animal issues is called the *mouth*, and its outer edge the *lip ;* each turn of the shell is a *whorl ;* the last and biggest, the *body-whorl* , the whorls, from the point at the

Fig. 8. *Spirula.*

top, or *apex*, down to the mouth form the *spire ;* and the line where the whorls join each other is called the *suture.*   The axis of the shell around which the whorls are coiled is sometimes open or hollow, and the shell is then said to be *umbilicated* (as in Fig. 3*b*) ; when closely coiled, a pillar of shell, or *columella*, is left (as in Fig. 9).   Sometimes the corner of the mouth farthest

from the spire and next the columella, is produced into a channel, the *anterior canal* (as in Fig. 9); whilst where the mouth meets the base of the spire there may be a kind of notch which is termed the *posterior canal*. Most Gasteropods are *dextral*, that is to say, the mouth is to the right of the axis as you look at it ; a few, however, are *sinistral*, or wound to the left (like *Physa*); whilst reversed varieties of both kinds are met with.

Gasteropods of the first order have comb-like gills placed in advance of the heart, and are hence termed PROSOBRANCHIATA. They are divided into two groups : (*a*) *Siphonostomata* (Tube-mouthed), in which the animal has a long proboscis, and a tube, or siphon, from the breathing-chamber that passes along the anterior canal of the shell, which in this group is well developed. They have a horny operculum, or lid, with which to close the aperture. (*b*) *Holostomata* (or Whole-mouthed). In these the siphon is not so produced, and does not want to be protected ; accordingly the mouth of the shell is *entire*, *i.e.* has no canal.

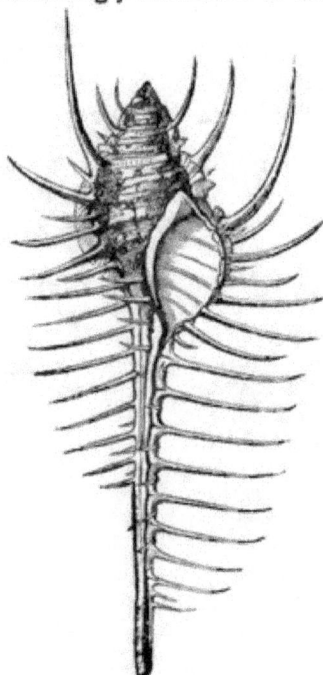

Fig. 9. *Murex tenuispina.*

The operculum is horny or shelly. The former (group *a*) includes several families :

1. *Strombidæ*, comprising shells, like the huge *Strombus*, or "Fountain-shell," which is so often used to adorn the mantel-piece or rockery, and from which cameos are cut.

2. The *Muricidæ*, of which the *Murex* (an extraordinary form of this is the "Venus' comb," *Murex tenuispina*, Fig. 9), the Mitre-shells, and the Red-Whelks(*Fusus*) are examples.

3. The *Buccinidæ*, taking its name from its type, the Common Whelk (*Buccinum undatum*), and including such other forms as the Dog-Whelk (*Nassa*), the *Purpura*, the strange *Magilus*, and the lovely Harp-Shells and Olives (Fig. 10).

4. The *Cassididæ*, or "Helmet-Shells." *Cassis rufa*, from West Africa, is noted as the best species of shell for cameo engraving ;

with it are classed the "Tun" (*Dolium*) and the great "Triton" (*Triton tritonis*), such as the sea-gods of mythology are represented blowing into by way of trumpet, and which are used by the Polynesian Islanders to this day instead of horns.

Fig. 10.
Oliva tessellata.

5. The *Conidæ*, whose type, the "Cone-shell" (Fig. 11), is at once distinctive and handsome, but which in the living state is covered by a dull yellowish-brown periostracum that has to be carefully removed before the full beauties of the shell are displayed.

6. The *Volutidæ*, embracing the Volutes and "Boat-shells" (*Cymba*).

7. The *Cypræidæ*, or Cowries (Fig. 12), which owe their high polish to the size of the shell-secreting organ (mantle), whose edges meet over the back of the shell, concealing it within its folds. With these is classed the "China-shell" (*Ovulum*).

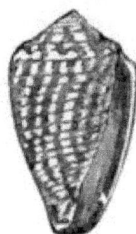

Fig. 11. *Conus vermiculatus.*

The second group, or *Holostomata*, is divided into nineteen families, beginning with—

1. The *Naticidæ*, whose type, the genus *Natica*, is well known to all shell-collectors through the common *Natica monilifera* of our coasts.

2. The *Cancellariadæ*, in which the shells are cancelled or cross-barred by a double series of lines running, one set with the whorls, and the other across them.

3. The *Pyramidellidæ*, which are high-spired, elongated, and slender shells, with the exception of the genus *Stylina*, which lives attached to the spines of sea-urchins or buried in living star-fishes and corals.

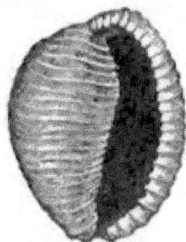

Fig. 12. *Cypræa oniscus.*

4. The *Solaridæ*, or "Staircase-shells," whose umbilicus is so wide that, as you look down it, the projecting edges of the whorls appear like a winding staircase. It is a very short-spired shell.

5. The *Scalaridæ*, "Wentle-traps" or "Ladder-shells," which may be readily recognised from their white and lustrous appearance and the strong rib-like markings of the periodic mouths that encircle the whorls.

6. The *Cerithiadæ*, or "Horn-shells," which are very high-spired, and whose columella and anterior canal are produced in the form of an impudent little tail, the effect of which, however, in the genus *Aporrhais*, or "Spout-shells," is taken away by the expanded and thickened lip.

7. In the next family, the *Turritellidæ*, or " Tower-shells," the type Turritella is spiral ; but in the allied form *Vermetus*, though the spire begins in the natural manner, it goes off into a twisted tube resembling somewhat an ill-made corkscrew. The mouth in this family is often nearly round.

8. The *Melaniadæ*, and 9, The *Paludinidæ*, are fresh-water shells. The former are turreted, and the latter conical or globular. Both are furnished with opercula, but the mouth in the first is more or less oval and frequently notched in front, while in the latter it is rounded and entire.

10. The *Litorinidæ*, or Periwinkles, need no word from us.

11. The *Calyptræidæ* comprise the "Bonnet-limpet," or *Pileopsis*, and "Cup-and-saucer-limpets" (*Calyptræa*). They may be described briefly as limpets with traces of a spire left. The genus *Phorus*, however, is spiral, and resembles a *Trochus.* They have been called "Carriers" from their strange habit of building any stray fragments of shell or stone into their house, thus rendering themselves almost indistinguishable from the ground on which they crawl.

12. The *Turbinidæ*, or "Top-shells," are next in order, and of these the great *Turbo marmoreus* is a well-known example, being prepared as an ornament for the whatnot or mantelpiece by removing the external layer of the shell in order to display the brilliant pearly nacre below. These mollusca close their mouths with a horny operculum, coated on its exterior by a thick layer of porcelain-like shelly matter. With them are classed the familiar *Trochus* and other closely allied genera.

13. The *Haliotidæ* offer in the representative genus *Haliotis*, or the "Ear-shell," another familiar mantelpiece ornament.

14. The *Ianthinidæ*, or "Violet-snails," that float about in mid-Atlantic upon the gulf-weed, and at certain seasons secrete a curious float or raft, to which their eggs are attached, are next in order, and are followed by—

15. The *Fissurellidæ*, or "Key-hole" and "Notched limpets," whose name sufficiently describes them. To these succeed—

16. The *Neritidæ*, an unmistakable group of globular shells, having next to no spire and a very glossy exterior, generally ornamented with a great variety of spots and bands.

17. The *Patellidæ*, or true Limpets, are well known to every sea-side visitor : large species, as much as two inches across, are to be found on the coast of Devon, but these are pigmies compared with a South American variety which attains a foot in diameter

18. The *Dentaliadæ*, represented by the genus *Dentalium*, or

'Tooth-shell," are simply slightly curved tubes, open at both ends and tapering from the mouth downwards, and cannot be mistaken.

19. Lastly, we have the *Chitonidæ*, whose single genus *Chiton* possesses shells differing from all other mollusca in being composed of eight plates overlapping each other, and in appearance reminding one of the wood-louse. This animal is not only like the limpet in form but also in habits, being found adhering to the rocks and stones at low-water.

Order II.—PULMONIFERA. Contains the air-breathing *Gasteropods*, and to it consequently belong all the terrestrial mollusca, though some few aquatic genera are also included. The members of this order have an air-chamber instead of gills, and are divided into two groups, (*a*) those without an operculum, and (*b*) those having an operculum. Foremost in the first group stands the great family—

1. *Helicidæ*, named after its chief representative, the genus *Helix*. It also includes the "Glass-shell" (*Vitrina*), the "Amber-shell (*Succinea*), and such genera as *Bulimus*, *Achatina*, *Pupa*, *Clausilia* (Fig. 13), etc., which differ from the typical *Helix* in appearance, possessing as they do comparatively high spires.

2. The *Limacidæ*, or "slugs," follow next; of these only one, the genus *Testacella*, has

Fig. 13. *Clausilia biplicata.*

an external shell stuck on the end of its tail; the rest have either a more or less imperfect shell concealed underneath the mantle, or else none at all.

3. The *Oncidiadæ* are slug-like, and devoid of shell.

4. The *Limnæidæ* embrace the "Pond-snails," chief of whom is the well-known, high-spired *Limnæa stagnalis*. Other shells of this family associated with *Limnæa* are, however, very different in shape; for instance, *Physa* has its whorls turning to the left instead of to the right; *Ancylus* (Fig. 24), or the fresh-water limpet, is of course limpet-like; while *Planorbis*, or the "Coil-shell," is wound like a watch-spring.

5. The *Auriculidæ* includes both spiral shells, such as *Auricula* and *Charychium*, and a limpet-like one *Siphonaria*.

At the head of group *b* stands 1, *Cyclostomidæ*. *Cyclostoma*

D

elegans is a common shell on our chalk-downs, and well illustrates its family, in which the mouth is nearly circular. Foreign examples of this genus are much esteemed by collectors. The other two families are, (2) *Helicinidæ* and (3) *Aciculidæ*.

Order III.—OPISTHOBRANCHIATA. These animals carry their gills exposed on the back and sides, towards the rear of the body. Only a few have any shell. 1. The *Tornatellidæ*, which have a stout little spiral shell. 2. The *Bullidæ*, in which the spire is concealed (Fig. 26). 3. The *Aplysiadæ*, where the shell is flat and oblong or triangular in shape. The remaining families are slug-like and shell-less.

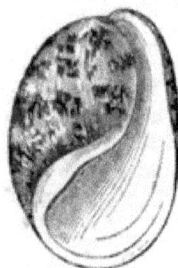

Order IV.—NUCLEOBRANCHIATA. Derives its name from the fact that the animals constituting it have their respiratory and digestive organs in a sort of nucleus on the posterior part of the back, and covered by a minute shell. As they are pelagic, the shells are not readily to be obtained. They are divided into two families, *Firolidæ* and *Atlantidæ*.

Fig. 14. *Bulla ampulla.*

CLASS III.—PTEROPODA. Like the last, these pretty little mollusca are ocean-swimmers. The members of one division of them, to which the *Cleodora* belongs, is furnished with iridescent external shells.

CLASS IV.—The LAMELLIBRANCHIATA (Plate-gilled), or CONCHIFERA (Shell-bearing), includes the mollusca commonly known as "bivalves," the animal being snugly hidden between two more or less closely fitting shelly valves. The oysters, cockles, etc, are examples of this class. The two valves are fastened together near their points, or beaks (technically called *umbones*), by a tough elastic ligament, sometimes supplemented by an internal cartilage. If this be severed and the valves parted, it will be found that in most cases they are further articulated by projecting ridges or points called the *teeth*, which, when the valves are together, interlock and form a hinge; the margin of the shell on which the teeth and ligament are situated is termed the *hinge-line*. A bivalve is said to be *equivalve* when the two shells composing it are of the same size, *inequivalve* when they are not. If the umbones are in the middle, the shell is *equilateral* (Fig. 15); but *inequilateral* when they are nearer one side than the other (Fig. 16). If the shell be an oyster or a scallop, you will find on the inside a single circular scar-like mark near the centre ; this is the point to which the muscles that close the valves and hold them so tightly together are attached. In the majority of bivalves, however, there are two such muscular impressions, or scars, one on either side of each valve of the shell. The former group on this account are often called *Monomyaria* (having one shell-muscle), and the latter *Dimyaria* (having two shell-muscles). In the last named the two muscular impressions are united by a fine groove (or *pallial-line*), which in some runs parallel to the margin of the shell (Fig. 15), whilst in others it makes a bend in (*pallial-sinus*) on one side of the valve towards the centre (Fig. 16). In Monomyaria it will be found running parallel to the margin of the shell. It marks the line of attachment of the mantle or

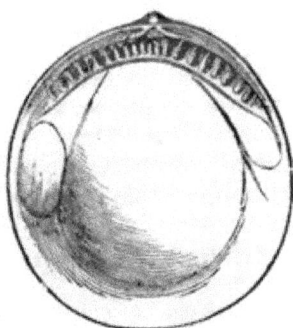

Fig. 15. *Petunculus Gurangeri.*

shell-secreting organ of the animal to the shell which grows by the addition of fresh matter along its edges, so that the concentric curved markings so often seen on the exterior correspond in their origin with the periodic mouths of the Gasteropods. The bivalves are all aquatic, and many bury themselves in the sand or mud by means of a fleshy,

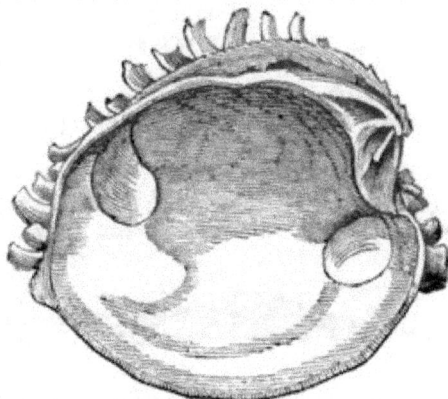

Fig. 16. *Venus plicata.*

muscular foot. These are furnished with two siphons, or fleshy tubes, sometimes united, sometimes separate, through which they respire, drawing the water in through one and expelling it by the other. Those kinds whose habit it is to bury themselves below the surface of the mud or sand are furnished with long retractile siphons, and to admit of their withdrawal into the shell, the mantle is at this point attached farther back, giving rise to the *pallial-sinus* above described ; this sinus is deeper as the siphons are proportionately longer, and in many cases, too, the valves do not meet at this point when the shell is closed.

Attention to these particulars is necessary when arranging your bivalves, as on them their classification depends, the class being divided into—

　a. ASIPHONIDA (Siphonless).

　b. SIPHONIDA *Integro-pallialia* (with Siphons).—Pallial-line entire.

　c. SIPHONIDA *Sinu-pallialia* (with Siphons).—Sinus in pallial-line.

DIVISION *a.*—ASIPHONIDA—is next subdivided into—

　1. The *Ostreidæ*, or oysters, which are deservedly a distinct family in themselves.

　2. The *Anomiadæ*, comprising the multiform and curiously constructed *Anomia*, with the " Window-shells " (*Placuna*).

　3. The *Pectinidæ*, taking its name from the genus *Pecten*, or " Scallop-shells," of which one kind (*P. maximus*) is frequently to be seen at the fishmongers' shops. The " Thorney oysters " (*Spondylus*) take rank here, and are highly esteemed by collectors, one specimen indeed having been valued at £25 !

　4. The *Aviculidæ*, or " Wing-shells," among which are numbered the " Pearl-oyster " of commerce (*Meleagrina margaritifera*). The strange T-shaped " Hammer oyster " belongs to this family, as does also the *Pinna*. The Pinnas, like the mussels and some other bivalves, moor themselves to rocks by means of a number of threads spun by the foot of the

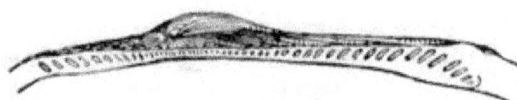

Fig. 17. Hinge-teeth of *Arca barbata*.

mollusc, and termed the *byssus*, which in this genus is finer, more silky, than in any other, and has been woven into articles of dress.

　5. The *Mytilidæ*, or mussels, including the *Lithodomus*, or " Date-shell," which bores into corals and even hard limestone rocks.

　6. The *Arcadæ*, or " Noah's-ark-shells," characterized by their long straight hinge-line set with numerous very fine teeth (Fig. 17). The " Nut-shell " (*Nucula*) belongs to this family.

7. The *Trigoniadæ*, whose single living genus, the handsome *Trigonia* (Fig. 18), is confined to the Australian coast-line, whereas in times now long past they had a world-wide distribution.

8. The *Unionidæ*, comprising the fresh-water mussels.

DIVISION *b.*—SIPHONIDA *Integro-pallialia.*

1. The *Chamidæ*, represented by the reef-dwelling *Chama*.

Fig.18. *Trigonia margaritacea.*

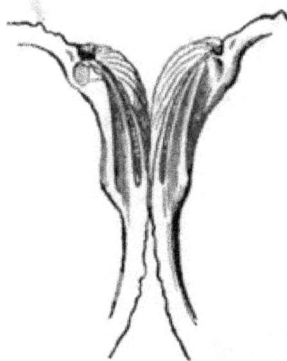

2. The *Tridacnidæ*, whose sole genus *Tridacna* contains the largest specimen of the whole class of bivalves, the shells sometimes measuring two feet and more across.

3. The *Cardiadæ*, or cockles.

4. The *Lucinidæ*, in which the valves are nearly circular, and as a rule not very attractive in appearance, though the " Basket-shell " (*Corbis*) has an elegantly sculptured exterior.

5. The *Cycladidæ*, whose typical genus *Cyclas*, with its round form and thin horny shell, is to be found in most of our ponds and streams.

6. The *Astartidæ*, a family of shells having very strongly developed teeth, and the surface of whose valves is often concentrically ribbed.

Fig. 19. Hinge of *Cardita sinuata.*

7. The *Cyprinidæ*, which have very solid oval or elongated shells and conspicuous teeth (Fig. 19). The " Heart-cockle " (*Isocardia*) belongs to this family.

DIVISION *c.*—SIPHONIDA *Sinu-pallialia.*

1. The *Veneridæ*. The hard, solid shells of this family are for elegance of form and beauty of colour amongst the most attractive a collector can possess. Their shells are more or less oval and have three teeth in each valve (Fig. 20).

Fig. 20. Hinge of *Cytherea crycina.*

2. The *Mactridæ* are somewhat triangular in shape, and may be at once recognised by the pit for the hinge-ligament, which also assumes that form, as seen in the accompanying figure of *Lutraria elliptica* (Fig. 21).

Fig. 21. Hinge of *Lutraria elliptica*

3. The *Tellinidæ* comprise some of the most delicately tinted, both externally and internally, of all shells. In some, coloured bands radiate from the umbones, and well bear out the fanciful name of "Sunset shells" bestowed upon them. Their valves are generally much compressed.

4. The *Solenidæ*, or "Razor-shells," rank next, and are readily recognised by the extreme length of the valves in proportion to their width, and also by their gaping at both ends.

5. The *Myacidæ*, or "Gapers," have the siphonal ends wide apart (in the genus *Mya* both ends gape), and are further characterized by the triangular process for the cartilage, which projects into the interior of the shell. One valve (the left) is generally smaller than the other.

6. The *Anatinidæ* have thin, often inequivalve pearly shells. The genus *Pandora* is the form most frequently met with in collections.

7. The *Gastrochænidæ* embraces two genera (*Gastrochæna* and *Saxicava*) of boring mollusca, which perforate shells and rocks, and also the remarkable tube-like "Watering-pot-shell" (*Aspergillum*) which is hardly recognisable as a bivalve at all.

8. The *Pholadidæ* concludes the list of bivalves, and comprises the common rock-boring Pholas (Fig. 22) of our coasts and the wood-boring ship-worm "Teredo" (Fig. 23).

Although the *Brachiopoda*, or "Lamp-shells," are not true mollusca, they are not very far removed from them, and are so often to be found in cabinets that it will not do to pass them over, especially since in past times they were very abundant, an enormous number occurring in the fossil state. Only eight genera are now living. Shells belonging to this class are readily recognised by the fact of one valve being larger than the other, and possessing a distinct

Fig. 22.
Pholas dactylus.

beak, the apex of which is perforated. The *Terebratulidæ* are the most extensive family of this class.

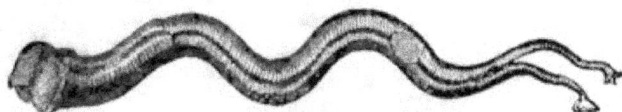

Fig. 23. *Teredo navalis.*

HOW TO ARRANGE THE SHELLS IN THE CABINET.

When you have arranged your specimens in the order above indicated, proceed to place them in your boxes, arranging and labelling them after the manner shown in the accompanying diagram.

Class.				
Order.	Species.	Species.	Species.	Species.
Family Name.				
Generic Name.	Species.	Species.	Species.	Species.
Species.	Species.	Species.	Family Name.	Species.
			Generic Name.	
Species.	Generic Name.	Species.	Species.	Species.
Species.	Species.	Species.	Species.	Generic Name.
Species.	Species.	Generic Name.	Species.	Species.
Species.	Species.	Species.	Species.	Species.

On the lid, or on a slip of paper or card placed at the head of your columns of trays, write the class and order, with its proper number (I., II., etc., as the case may be) ; then at the top of your left-hand column place the family and its number, and under it the name of the first genus. The species (one in each tray) come next, then the name of the next genus following it, succeeded by its species, and so on.

The object of the young collector should be to obtain examples of as many *genera* as possible, since a collection in which a great number of genera are represented is far more useful and instructive than one composed of a great many species referable to but few genera. He will also find it very convenient to separate the British Shells from his general collection, sub-dividing them for convenience into "Land and Fresh-water," and "Marine." Of these he should endeavour to get every species, and even variety, making the thing as complete as possible. Or a separate collection may be made of all those kinds which he can find within a certain distance of his own home. A collection of this sort possesses, in addition to its scientific worth, an interest of tis own, owing to the local associations that invariably connect themselves with it.

TABLE OF SOME OF THE MORE IMPORTANT GENERA, SHOWING THE APPROXIMATE NUMBER OF SPECIES BELONGING TO EACH GENUS AND THEIR DISTRIBUTION.

CLASS I.—Cephalopoda.

Order I.—Dibranchiata.

Section A.—*Octopoda.*

Family.	Genus.	No. of Species.	Distribution.
1.	Argonauta	4	Tropical seas.
2.	Octopus	46	Rocky coasts in temperate and tropical regions.

Section B.—*Decapoda.*

3.	Loligo	19	Cosmopolitan.
4.	Sepia	30	On all coasts.
5.	Spirula	3	All the warmer seas.

Order II.—Tetrabranchiata.

6	Nautilus	3 or 4	Chinese Seas, Indian Ocean, Persian Gulf.

CLASS II.—Gasteropoda.

Order I.—Prosobranchiata.

Division *a.—Siphonostomata.*

Family.	Genus.	No. of Species.	Distribution.
1.	Strombus	60	W. Indies, Mediterranean, Red Sea, Indian Ocean, Pacific—low water to 10 fathoms.
	Pteroceras	12	India, China.
2.	Murex	180	On all coasts.
	Columbella	200	Sub-tropical regions, in shallow water on stones.
	Mitra	350	Tropical regions, from low water to 80 fathoms.
	Fusus	100	On all coasts.
3.	Buccinum	20	Northern seas, from low water to 140 fathoms.
	Eburna	9	Red Sea, India, Australia, China, Cape of Good Hope.
	Nassa	210	World-wide—low water to 50 fathoms.
	Purpura	140	World-wide—low water to 25 fathoms.
	Harpa	9	Tropical—deep water, sand, muddy bottoms.
	Oliva	117	Sub-tropical—low water to 25 fathoms.
4.	Cassis	34	Tropical regions, in shallow water.
	Dolium	15	Mediterranean, India, China, W. Indies, Brazil, New Guinea, Pacific.
	Triton	100	Temperate and sub-tropical regions, from low water to 50 fathoms.
	Ranella	50	Tropical regions, on rocks and coral-reefs.
	Pyrula	40	Sub-tropical regions, in 17 to 35 fathoms.
5.	Conus	300	Equatorial seas—shallow water to 50 fathoms.
	Pleurotoma	500	Almost world-wide—low water to 100 fathoms.
6.	Voluta	100	On tropical coasts, from the shore to 100 fathoms.
	Cymba	10	West Coast of Africa, Lisbon, Straits of Gibraltar.
	Marginella	90	Mostly tropical.
7.	Cypræa	150	Warmer seas of the globe, on rocks and coral-reefs.
	Ovulum	36	Britain, Mediterranean, W. Indies, China, W. America.

Division *b.—Holostomata.*

Family.	Genus.	No. of Species.	Distribution.
8.	Natica	90	Arctic to tropical regions, on sandy and gravelly bottoms, from low water to 90 feet.
	Sigaretus	26	E. and W. Indies, China, Peru.
9.	Cancellaria	70	W. Indies, China, S. America, E. Archipelago—low water to 40 fathoms.
10.	Pyramidella	11	W. Indies, Mauritius, Australia, in sandy bays and on shallow mud-banks.
	Odostomia	35	Britain, Mediterranean, and Madeira—low water to 50 fathoms.
	Chemnitzia	70	World-wide—low water to 100 fathoms.
	Eulima	26	Cuba, Norway, Britain, India, Mediterranean, Australia—5 to 90 fathoms
11.	Solarium	25	Sub-tropical and tropical—widely distributed.
12.	Scalaria	100	World-wide—low water to 100 fathoms.
13.	Cerithium	100	World-wide.
	Potamides	41	Africa and India, in mud of large rivers.
	Aporrhais	3	Labrador, Norway, Britain, Mediterranean—20 to 100 fathoms.

Family.	Genus.	No. of Species.	Distribution.
14.	Turritella	50	World-wide—low water to 100 fathoms.
	Vermetus	31	Portugal, Mediterranean, Africa, India.
15.	Melania	160	S. Europe, India, Philippines and Pacific Islands—in rivers.
	Melanopsis	20	Spain, Australia, Asia Minor, New Zealand—in rivers.
16.	Paludina	60	Northern Hemispheres, Africa, India, China, etc.—in lakes and rivers.
	Ampullaria	50	S. America, W. Indies, Africa, India—in lakes and rivers.
17.	Litorina	40	On all shores.
	Rissoa	70	World-wide—in shallow water on sea-weed to 100 fathoms.
18.	Calyptrea	50	World-wide—adherent to rocks, etc.
	Crepidula	40	West Indies, Mediterranean, Cape of Good Hope, Australia.
	Pileopsis	7	Britain, Norway, Mediterranean, E. and W. Indies, Australia.
	Hipponyx	70	W. Indies, Galapagos, Philippines, Australia.
	Phorus	9	W. Indies, India, Javan and Chinese Seas—in deep water.
19.	Turbo	60	On the shores of Tropical seas.
	Phasinella	30	Australia, Pacific, W. Indies, Mediterranean.
	Imperator	20	S. Africa, India, etc.
	Trochus	150	World-wide—from low water to 100 fathoms.
	Rotella	18	India, Philippines, China, New Zealand.
	Stomatella	20	Cape, India, Australia, etc.
20.	Haliotis	75	Britain, Canaries, India, Australia, California—on rocks at low water.
	Stomatia	12	Java, Philippines, Pacific, etc.—under stones at low water.
21.	Ianthina	6	Gregarious in the open seas of the Atlantic and Pacific.
22.	Fissurella	120	World-wide—on rocks from low water to 5 fathoms.
	Emarginula	26	Britain, Norway, Philippines, Australia—from low water to 90 fathoms.
23.	Nerita	116	On the shores of all warm seas.
	Neritina	110	In fresh waters of all warm countries, and in Britain.
	Navicella	24	India, Mauritius, Moluccas, Australia, Pacific—in fresh water, attached to stones.
24.	Patella	100	On all coasts—adhering to stones and rocks.
25.	Dentalium	30	World-wide—buried in mud.
26.	Chiton	200	World-wide—low water to 100 fathoms.

ORDER II.—Pulmonifera.

Division *a.—Inoperculata.*

27.	Helix	1,600	World-wide—on land in moist places.
	Succinea	68	
	Bulimus	650	
	Achatina	120	World-wide—burrowing at roots and bulbs.
	Pupa	236	World-wide—amongst wet moss.
	Clausilia	400	Europe and Asia—in moist spots.

Family.	Genus.	No. of Species.	Distribution.
28.	Limax	22	Europe and Canaries—on land in damp localities.
	Testacella	3	S. Europe, Canaries, and Britain—burrowing in gardens.
29.	Oncidium	16	Britain, Red Sea, Mediterranean -on rocks on the seashore.
30.	Limnæa	50	Europe, Madeira, India, China, N. America—in ponds, rivers, lakes, etc.
	Physa	20	America, Europe, S. Africa, India, Philippines—in ponds, rivers, lakes, etc.
	Ancylus	14	Europe, N. and S. America—in ponds, rivers, lakes, etc.
	Planorbis	145	Europe, N. America, India, China—in ponds, rivers, lakes, etc.
31.	Auricula	50	Tropical—in salt marshes.
	Siphonaria	30	World-wide—between high and low water.

Division b.—Operculata.

32.	Cyclostoma	80	S. Europe, Africa ⎫
	Cyclophorus	100	India, Philippines ⎬—on land.
	Pupina	80	Philippines, New Guinea ⎭
33.	Helicina	150	W. Indies, Philippines, Central America, Islands in Pacific—on land.
34.	Acicula	5	Britain, Europe, Vanicoro—on leaves and at roots of grass.
	Geomelania	21	Jamaica—on land.

ORDER III.—Opisthobranchiata.

Division a.—Tectibranchiata.

35.	Tornatella	16	Red Sea, Philippines, Japan—in deep water.
36.	Bulla	50	Widely distributed—low water to 30 fathoms.
37.	Aplysia	40	Britain, Norway, W. Indies—low water to 15 fathoms, on seaweed.
38.	Pleurobranchus	20	Britain, Norway, Mediterranean.

Division b.—Nudibranchiata.

39-44.		All shell-less.

ORDER IV.—Nucleobranchiata.

45.	Firola	8	Atlantic, Mediterranean.
	Carinaria	5	Atlantic and Indian Oceans.
46.	Atlanta	15	Warmer parts of the Atlantic.

CLASS III.—PTEROPODA.

Division *a.—Thecosomata.*

Family	Genus.	No. of Species.	Distribution.
1.	Hyalea	19	Atlantic, Mediterranean, Indian Ocean.
	Cleodora	12	
2.	Limacina	2	Arctic and Antarctic Seas.

Division *b.—Gymnosomata.*

3.	Clio, etc.		Shell-less.

CLASS IV.—LAMELLIBRANCHIATA.

Division *a.—Asiphonida.*

1.	Ostrea	100	World-wide—in estuaries, attached.
2.	Anomia	20	India, Australia, China, Ceylon—attached to shells from low water to 100 fathoms.
	Placuna	4	Scinde, North Australia, China—in brackish water.
3.	Pecten	176	World-wide—from 3 to 40 fathoms.
	Lima	20	Norway, Britain, India, Australia—from 1 to 150 fathoms.
	Spondylus	70	Tropical seas—attached to coral-reefs.
4.	Avicula	25	Britain, Mediterranean, India—25 fathoms.
	Perna	18	In tropical seas—attached.
	Pinna	30	United States, Britain, Mediterranean, Australia, Pacific—low water to 60 fathoms.
5.	Mytilus	70	World-wide—between high and low water mark.
	Modiola	70	British and tropical seas—low water to 100 fathoms.
6.	Arca	400	In warm seas—from low water to 200 fathoms.
	Pectunculus	58	West Indies, Britain, New Zealand—from 8 to 60 fathoms.
	Nucula	70	Norway, Japan—from 5 to 100 fathoms.
7.	Trigonia	3	Off the coast of Australia.
8.	Unio	420	World-wide—in fresh waters.
	Anodon	100	North America, Europe, Siberia—in fresh waters.

Division *b.—Siphonida.*

9.	Chama	50	In tropical seas on coral reefs.
10.	Tridacna	7	Indian and Pacific Oceans, Chinese Seas.
11.	Cardium	200	World-wide—from the shore line to 140 fathoms.
12.	Lucina	70	Tropical and temperate seas—sandy and muddy bottoms—from low water to 200 fathoms.
	Keflia	20	Norway, New Zealand, California—low water to 200 fathoms.
13.	Cyclas	60	Temperate regions—in all fresh waters.
	Cyrena	130	From the Nile and other rivers to China—and in mangrove swamps.
14.	Astarte	20	Mostly Arctic—from 30 to 112 fathoms.
	Crassatella	34	Australia, Philippines, Africa, etc.

Family.	Genus.	No. of Species.	Distribution.
15.	Cyprina	1	From Britain to the most northerly point yet reached —from 5 to 80 fathoms.
	Circe	40	Britain, Australia, India, Red Sea—8 to 50 fathoms.
	Isocardia	5	Mediterranean, China, Japan—burrowing in sand.
	Cardita	54	Tropical seas—from shallow water to 150 fathoms.
16.	Venus	176	⎫ World-wide—buried in sand, from low water to 100
	Cytherea	113	⎭ fathoms.
	Artemis	100	Northern to tropical seas—from low water to 100 fathoms.
	Tapes	80	Widely distributed—burrowing in sand, from low water to 100 fathoms.
	Venerupis	20	Britain, Canaries, India, Peru—in crevices of rocks.
17.	Mactra	125	World-wide—burrowing in sand.
	Lutraria	18	Widely distributed—burrowing in sand.
18.	Tellina	300	In all seas—from the shore line to 15 fathoms.
	Psammobia	50	Britain, Pacific and Indian Oceans—from the littoral zone to 100 fathoms.
	Sanguinolaria	20	W. Indies, Australia, Peru.
	Semele	60	Brazil, India, China, etc.
	Donax	68	Norway, Baltic, Britain—in sand near low water mark.
19.	Solen	33	World-wide—burrowing in sand.
	Solecurtus	25	Britain, Africa, Madeira, Mediterranean—burrowing in sand.
20.	Mya	10	North Seas, W. Africa, Philippines, etc.—river mouths from low water to 25 fathoms.
	Corbula	60	United States, Britain, Norway, Mediterranean, W. Africa, China—from 15 to 80 fathoms.
21.	Anatina	50	India, W. Africa, Philippines, New Zealand.
	Thracia	17	Greenland to Canaries and China—from 4 to 120 fathoms.
	Pandora	18	Spitzbergen, Panama, India—from 4 to 110 fathoms, burrowing in sand and mud.
22.	Gastrochæna	10	W. Indies, Britain, Red Sea, Pacific Islands—from shore line to 30 fathoms.
	Saxicava		Arctic Seas, Britain, Mediterranean, Canaries and the Cape—in crevices and boring into limestone and rocks.
	Aspergillum	21	Red Sea, Java, New Zealand—in sand.
23.	Pholas	32	Almost universal—from low water to 25 fathoms.
	Xylophaga	2	Norway, Britain, S. America—boring into floating wood.
	Teredo	14	In tropical seas—from low water to 100 fathoms.

SOME WORKS OF REFERENCE.

MOLLUSCA IN GENERAL.

"A Manual of Mollusca." By Dr. S. P. Woodward.
"Tabular View of the Orders and Families of the Mollusca."
Published by the Society for Promoting Christian Knowledge.
"Cassell's Natural History," latest edition, article on the
Mollusca. By Dr. Henry Woodward.

BRITISH MOLLUSCA.

"A History of British Mollusca and their Shells." By
Professor E. Forbes and S. Hanley.
"British Conchology." By J. G. Jeffreys.
"Common Shells of the Sea-shore." By Rev. J. G. Wood.

BRITISH LAND AND FRESH-WATER MOLLUSCA.

"Land and Fresh-water Mollusca indigenous to the British
Isles." By Lovell Reeve.
"A Plain and Easy Account of the Land and Fresh-water
Mollusca of Great Britain." By Ralph Tate.

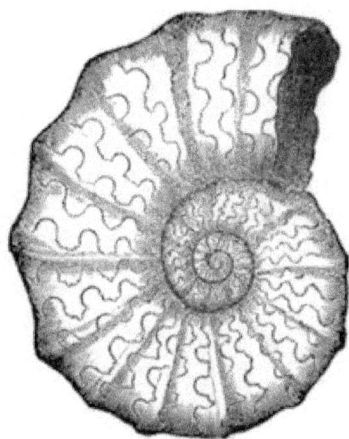

FOSSILS.

BY

B. B. WOODWARD.

FOSSILS.

INTRODUCTORY.

GEOLOGY is of all "hobbies" the one best calculated not only to develop the physical powers, but also, if pursued with any degree of earnestness, to train and extend the mental faculties. To study geology properly, the rocks themselves must be visited and carefully observed, their appearance noted, and the fossils, if any, which they contain, collected. This necessitates many a pleasant walk into the open country to quarries and cuttings, or rambles along the sea-shore to cliffs which may be worth investigating, whilst botany, entomology, or any other congenial pursuit, may be followed on the way ; for natural science in its different branches has so many points of connection that it is impossible to study one of them without increasing one's interest in, and knowledge of, all the others. Again, in arranging, classifying, and studying at home the specimens collected on these expeditions, many an hour may be usefully spent ; habits of exactitude and neatness are acquired ; whilst in endeavouring to draw correct conclusions as to the way in which particular rocks were formed, and by what agencies brought to their present position, the reasoning faculties are exercised and developed.

The existence of fossil shells and bones in various strata of the earth's crust attracted attention at a very early date of the world's history ; the Egyptian priests were aware of the existence of marine shells in the hills bounding the Nile valley, and from this fact Herodotus drew the conclusion that the sea formerly covered the spot. The bones of the larger mammalia (rhinoceros, elephant, etc.), were, however, thought by the ancients to be human, and hence arose the idea of a race of giants having existed at some previous period of the earth's

history.   It was not, however, until near the end of the last
century that geology began to be recognised as a science, and
the true bearing of fossils in relation to the rocks in which they
were found was conclusively proved.   William Smith in England, and Werner in Germany, while working independently of
each other, both came to the same conclusion, viz. that the numerous strata invariably rested on each other in a certain order,
and that this order was never inverted,[1] whilst William Smith in
addition proved that each group of rocks, and even each stratum,
had its own peculiar set of fossils, by which it might be recognised
wherever it occurred.   From that time forth the study of the
various fossils began to be considered as a separate science
apart from that of the beds containing them ; this is now known
as Palæontology, the study of the composition of the rocks themselves being termed Petrology.

At this moment, however, we are less concerned with the
study of rocks and fossils than with the best and simplest way
of collecting, preparing, and arranging specimens as a means to
this study.

THE CABINET.

With regard to the cabinet for such specimens as you are
able to collect, the same advice holds good as that given
in a previous Manual (The Young Collector's Shell Book),
namely, the simpler the cabinet the better, though of course
card-board boxes would not as a rule be strong enough to stand
the weight of the specimens, and hence it is advisable to have
wooden ones.   The boxes in which Oakey's Wellington Knife-
powder is sent out (they measure about 15 in. × 10 in. × 3
in.) are on the whole the most convenient size, and are easily
obtainable at any oil and colourman's.   These, when painted
over with Berlin Black, after first removing the external labels,
look very neat.   The inside may be papered according to taste,
when the trays may be arranged in order ready for the reception
of your specimens.[2]

IMPLEMENTS REQUIRED WHEN COLLECTING.

A certain amount of apparatus is needful in collecting geological specimens.   It is necessary to break open the hard

---

[1] Except in such cases where the rocks themselves have been displaced
by movements of the earth's crust.
[2] For description of trays, see " The Young Collector's Shell-Book."

rocks in order to get at the fossils within, and for this purpose a strong hammer is required. One end of the hammer-head should be square, tapering, slightly, to a flat striking face; for when thus shaped the edges and corners are less likely to break off; the other side should be produced into a rather long, flat, and slightly curved pick, terminating in a chisel-edge at right-angles to the handle; the total length of the head should not exceed 9½ in., the striking face being 3 in. from the centre of the eye in which the handle (18 in. long) is inserted; the latter should be made of the toughest ash, American hickory, or "green-heart," and fixed in with an iron wedge ("roughed" to prevent its coming out again), taking care that ¼ in. of the handle protrudes on the other side. It is the usual practice, but a mistaken one, to cut it off level with the hammer head, which is likely, under these circumstances, to come off after it has been in use for a time, whereas, by leaving a small portion of the wedged-out end projecting, this mischance is avoided, and your weapon will not fail even when used to drag its owner up a stiff ascent. It is better to shape and fix the handle yourself, as by this means you can not only cut it to fit your hand, but may rely upon its being properly fastened in. By filing grooves around it an inch apart, it will serve to take rough measurements with, while a firm grasp may be insured by bees-waxing instead of polishing it. Another and much smaller hammer will also be necessary, chiefly for home use, to trim the specimens before putting them away in the cabinet; the head of this hammer must not be more than 2½ inches long, the handle springing from the centre; one end has a flat striking face, square in section, the other, instead of being formed like a pick, is wedge-shaped, the sharp edge being at right-angles to the handle. Next to a hammer, a cold chisel is indispensable to the collector, since without its aid many a choice specimen embedded in the middle of a mass of rock too large to break with the hammer would have to be left behind. There is one thing, however, to beware of in using this tool—it has sometimes to be hit with great force, and should you chance to miss it and strike your hand instead, the result may be more serious than even a severe bruise. To prevent this, procure from the shoemaker or saddler a piece of thick leather, about 4 inches in diameter, having a hole cut in the centre through which to pass the shank of the chisel, and, thus protected, you may wield the hammer with impunity.

For digging fossils out of clay, an old, stout knife, such as the worn-down stump of a carver, is handy, and in sandy beds an ordinary garden trowel is very useful, whilst in a chalk-pit a small saw is sometimes of great aid in extricating a desirable specimen.   The same may be said of an ordinary carpenter's wood-chisel.   For picking up small and delicate specimens, a pair of forceps should be carried, whilst without a pocket lens no true naturalist ever stirs abroad.   An ordinary stout canvas satchel, such as is commonly used by schoolboys, is the best thing for carrying home your specimens ; this may be made much stronger by the addition of two short strips of leather stitched on the back and running, one from each ring, to which the strap passing over the shoulder is fastened, down to the bottom of the bag ; by leaving a small portion unstitched near the bottom of each of these, wide enough for the shoulder-strap to pass through, the satchel may at a moment's notice be slung knapsackwise on the shoulders—a method of carrying it which is, as all who have tried it know, by far the most convenient when it is heavily laden or not in immediate requisition. A stout leather belt may be worn in which to carry all your hammers, supporting it on the side where the heavy hammer hangs by a band passing over the opposite shoulder. Before starting on an excursion, make a practice of seeing that you have everything with you, or when the critical moment comes, and some choice and fragile specimen is ready to be borne off, you may find that you are without the means necessary for taking it home. For ordinary hard specimens, newspaper well crumpled around them is without its equal, but some of the more delicate must be first wrapped in tissue paper or even cotton-wool, whilst the most fragile fossils should be packed in tins with bran or sawdust, the particles of which fill in all the corners and press equally everywhere, a useful faculty which cotton wool does not possess. When neither of these are to be obtained, dry sand will answer quite as well, though it is heavier to carry.

Although not absolutely necessary in the field, it is often useful to have a small bottle of acid in your pocket (nitric acid diluted to 1-12th with distilled water is the best) with which to test for limestones ; a drop of acid placed on a rock will, if there be any carbonate of lime in it, immediately begin to fizz.   Finally, every young collector should carry a note-book, and carefully record in it what he sees in each pit he visits, while, if it can be procured or borrowed, a geological map of the district you are exploring is a great help, for with its aid and that of a good compass you become practically independent of much extraneous assistance.

HOW TO USE YOUR IMPLEMENTS.

We will suppose by way of illustration that near us flows a river, on the rising ground above which is a pit that we propose to visit for the purpose of putting our apparatus into practical operation. When we have reached the floor of the pit, and stand looking up at the section before us, we are at first rather puzzled as to what the beds, which we see before us, are; for as the pit has not been worked for some time, its sides are partially overgrown with grass, and in places bits and pieces of the upper beds have fallen down and form a heap beneath which the lower ones lie buried. We must therefore make our way to those spots where the beds are left clear, and find out, if possible, what they are. By climbing up one of the heaps of fallen earth (*talus*) we reach the top, where, first of all, under the roots of the grass and shrubs, we find the mould in which these grow, and which is formed of the broken up (*disintegrated*) rocks forming the still higher ground above, and which the rains, frosts and snows, aided afterwards by the earthworms, have converted into mould. This, geologically speaking, is called *surface soil*, and is here about two feet deep. Just below it we find a layer of coarse gravel; the pebbles of which this is composed are of all sorts, sizes, and shapes, and are stained a deep brown by oxide of iron. Most of them are flints, and by diligent search you may find casts and impressions in these of sponges, shells, spines of sea urchins, etc. Flints, whether from gravel or their parent rock the chalk, are easiest broken by a light smart tap of the hammer, though when it is desired to shape them for the cabinet a soft iron hammer should be used, and the piece to be shaped placed on a soft pad on the knee, for when struck with a steel hammer flints splinter in all directions, and often through the very portion you most desire to preserve. In one spot we find a mass of sand included in the gravel; this mass is thickest in the middle, and tapers away towards each end, its total length being about fifty feet. Could we see the whole mass, we should probably find it to be a patch lying on the gravel and thinning out all around its edges; in other words it would be shaped like a lens—"*lenticular*" as geologists term it. When we examine this mass more closely, we find that the layers of sand do not run parallel with the bed, but are inclined in different directions, sometimes lying one way, sometimes another. This *false bedding* is due to the sand having been thrown down in waters agitated by strong currents that swept over the spot, now in one direction and now in another, scattering at one moment half the sand they had just piled up one way only to redeposit it the next minute in another. In the gravel also may be observed a similar though less marked arrangement, owing to the larger size of its constituents, which of course required a still stronger current action to wash them down.

Amongst the sand we now see some shells, and set to work to dig them out very carefully, for they are exceedingly brittle. The best specimens are to be obtained by throwing down masses of the sandy material and searching in it ; but only the stronger and finer examples will bear such usage. We next notice that these shells are precisely similar to those still found with living occupants in the river below, only they are no longer of a brownish colour, but owing to the loss of the animal matter of the shell have an earthy, dirty-white appearance. To carry these home they should be packed in bran in one of your tins with a note as follows made on a piece of paper and placed just inside—" Sand in gravel : topmost bed —— pit, August 2nd, 188-." Then if you are not able to work them out at once on reaching home, you will not forget whence they came. From the appearance of these sands and gravels, and the presence in them of shells exactly like those in the river below, it may reasonably be inferred that they once formed a portion of the bed of that river long ago, before it had scooped out its valley to the present depth. There is, however, something else in this sand-bed—a piece of bone protruding ; clear away the sand above it, and dig back until the whole is visible. It is broken through in one or two places, but otherwise is in fair condition ; remove the pieces carefully one by one, and wrap them in separate pieces of paper, and then proceed to search for others. These bones, which are plentiful in some of our river valley gravel-beds, are the remains of animals that once roamed in the forests which at that time covered the country ; they were probably either drowned in crossing the water, or got stuck in the mud on the banks on coming down to drink. A fine collection was made at Ilford by the late Sir Antonio Brady, and is now in the British Museum (Natural History) at South Kensington. Besides the bones of animals, you may expect to find examples of all, or nearly all, the different rocks in which the river has cut its valley, and samples of these may be picked out and taken home. Each specimen should be wrapped in a separate piece of paper to prevent its rubbing against others, care being taken to note the locality either by writing it on the paper or by affixing to the specimen a number corresponding to one in your note book against the description you have written of the bed. The gravel, with its accompanying bed of sand, may be traced down, by scraping away the surface, for about ten feet, when you will discover that it rests unevenly upon the beds below, which, instead of being horizontal, slope (*dip*) in a N.N.E. direction, making an angle of about 45° with the floor of the pit ; the gravel therefore rests successively upon the upturned ends of the lower beds, and, geologically speaking, is "unconformable" to them. Now as these underlying rocks were of course originally deposited in an horizontal position, they must have been pushed up and the upper parts worn away (*denuded*) before the gravel was deposited on them, for the accomplishment of which process an amount of time must have elapsed that it would be impossible to reckon by years.

When we come to examine these lower beds, we find first a stratum of stiff dark-brown clay containing fossils disposed in layers : those near the outer surface have been rendered so brittle by the weather, that it is necessary to make use of the pick end of the hammer and dig a little way into the face of the section before we come upon some which will bear removal by cutting them out with a knife. Pack them in a tin with bran, or, where much clay still adheres to them, wrap them in paper.

The true top of this bed is not visible, being concealed beneath a heap of earth in the corner of the pit, but we can see and measure about six feet of it.

The next bed in order is a light brownish band of sandy clay that splits along its layers into thin pieces or "*laminæ*," whence we may describe it as a sandy, *laminated* clay. On the freshly split surface of one piece we see scattered a number of small darker brown fragments ; an examination with a pocket lens clearly shows that these are little bits of leaves and stems, with here and there a more perfect specimen. These beds must have been deposited in the still waters just off the main stream of a large river which brought the plants floating down to this spot, where they became water-logged and sunk ; so, too, if you examine the shells in the bed immediately above, you will see that they are very like though not the same as those which at the present day love to dwell in the mud off the estuaries of big rivers in warmer parts of the globe ; hence we discover that at some far distant period a big river, but one which had no connection with that running close by, once flowed over this very spot. On tracing the leaf-bed down, we come all at once, at about three feet from its upper surface, upon a narrow band one or two inches thick of a substance composed of numerous bits of sticks and stalks closely matted together and partially mineralized. Vegetable matter in this form is known as lignite, and is one of the first stages towards the formation of coal out of plant remains. Below this lignite band we find our leaf-bed getting sandier and sandier, and losing all trace of the plants by degrees till it becomes almost pure sand. Here and there, however, it contains some curiously shaped masses, which, when broken through with the hammer, seem composed of nothing but the same grains of sand cemented together into a hard mass. In one there is, however, a curiously shaped hollow, which, upon examining it closely, you will see is a perfect cast of a small shell that has itself disappeared. A drop of acid on it fizzes away and sinks in between the grains of sand which in this spot become loose. A mass of sand or particles of clay thus cemented together, be it by iron, lime, or any other substance, is termed a "*nodule*" or "*concretion*," and in this particular instance has been formed as follows :—The rain-water falling on the sand where it comes to the surface sinks in and filters through the bed. Now there is always a certain amount of carbonic acid in rain-water, and this acid acted on the carbonate of lime of which the shell was composed, dissolving and dispersing it amongst the neighbouring grains of sand where it was re-deposited, cementing them together as we have seen. The bottom of this bed of sand we find to be just fifteen feet from the lignite band when measured at right-angles to the bed, and it is succeeded by a hard greyish rock, which requires a smart blow of the hammer to break it, but the surface of which, where it has been exposed to the weather, is much crumbled ("*weathered*"), and breaks readily into small pieces. It is easily scratched with the point of a knife, and therefore is not flint ; moreover, it fizzes strongly when touched with acid —hence there is a great deal of carbonate of lime in it, and we know that it is limestone.

Limestones are very largely, sometimes almost entirely, made up of the calcareous portions of marine creatures, such as the hard parts of corals, the tests of sea-urchins, the shells of mollusca, etc., welded, so to speak, into one mass by the heat, pressure, and chemical changes which the bed has undergone since its deposition at the bottom of the sea. There would be every reason, therefore, one might suppose, to expect a number of fossils in this bed ; but, alas ! disappointment awaits the young explorer, for with the exception of chalk and a few other limestones, these rocks are generally of such uniform texture that on being struck with the hammer they split through fossils and all, the fractured surface only too frequently showing nought save a few obscure markings. But what we fail to accomplish in our impatience, nature effects by slow degrees, and if you will turn over the weathered pieces and blocks lying about, you will soon find plenty of fossils sticking out all over them ; by a judicious use of hammer and chisel any of these may be detached and added to your stock, each being separately packed in paper and the locality written on the outside. Some seventy or eighty feet is all that is visible of this limestone ; the rest is unexcavated.

Before leaving the pit, it will be as well to select such rock specimens as you wish to place in your cabinet, trimming them to the required size on the spot, for should you, as is not unlikely, spoil two or three, you can readily pick a fresh one. Having secured our specimens, we will take a look at our note-book, to see if we have noted all the details we require. If so, our entries should run something as follows :—First, we have made a rough sketch of the position of the beds, carefully numbering each one ; then follow our notes on the individual beds, preceded by numbers corresponding with those in the sketch, thus :—

1. Surface Soil . . . . 2 ft.
2. River Gravel, including a lenticular mass of } 10 ft.
3. Sand, with land and fresh-water shells and bones of animals . }
4. Stiff dark-brown clay, with estuarine shells . . . 6 ft. seen.
5. Light-brown sandy clay, with leaves and stems of plants . . 3 ft.
6. Band of Lignite . . . . 2 in.
7. Same as 5, passing into— }
8. Pure Sand, with layers of concretions containing casts of shells } 15 ft.
9. Dark-Grey Limestone, with numerous fossils . . . 80 ft. seen.
  Beds 4 to 9 dip at an angle of 45° to the N.N.E.

Our imaginary pit is of course only a sort of geological Juan Fernandez, but it will serve in some degree to illustrate the method of dealing with various rocks and fossils when met with in the field, and how they may best be collected and carried home. A few additional suggestions where to look for fossils may, however, be given here. To begn with, inever neglect to search the fallen masses, especially their weathered surfaces, or to look carefully over the heaps of quarried materials, whatever they may happen to be, piled on the floor of the pit. In working at the beds themselves, remember that fossils frequently occur in layers which of course represent the old sea-bottom of the period ; to find these, it is necessary to follow the beds in a direction at right angles to their stratification, till you arrive at the sought-for layers, or *zones*.

Do not be surprised, when collecting from a formation you have never before studied, if the fossils are not at first apparent, though many are known to be present. The eye requires a few days in which to become accustomed to its fresh surroundings, and when the same spot has been carefully hunted over every day for a week, it is astonishing what a quantity of fossils are discernible where not one in the first instance was recognised.

HOW TO PREPARE THE SPECIMENS FOR THE CABINET.

The first thing to be done on unpacking our specimens is to pick out those which require the least attention, and get them out of the way. These will be your rock specimens, which, if they have been trimmed properly in the pit, will not need much further manipulation ; a word or two, however, as to the best method of proceeding when it is desirable to reduce a specimen, will not be out of place. If you wish to divide it in two, or detach any considerable portion, the specimen may, while held in the hand, be struck a smart blow with the hammer ; as, however, it not frequently happens that even with the greatest care the specimen under this treatment breaks in an opposite direction to that required, it is advisable to adopt a somewhat surer method, namely, to procure a block of tough wood, and in the centre bore a hole just large enough to receive the shank of the cold chisel, and thus hold it in an upright position with the cutting edge uppermost ; placing the specimen on this, and then hitting it immediately above with the hammer, it may be fractured through in any required direction. To trim off a small projection, hold the specimen in your hand with the corner towards you and directed slightly downwards, then with the edge of the striking face of the hammer hit it a smart blow at the line along which you wish it to break off ; the object of inclining the specimen is to make sure that the blow shall fall in a direction inclined away from the portion you wish to preserve, a *modus operandi* which it is necessary to bear well in mind if you would not spoil many a choice specimen. Anything beyond very general directions, however, it is impossible to give in such matters as this : experience, and a few hints from those who have themselves had practice in collecting and arranging specimens, are worth more than any written description, however lengthy and elaborate.

Having reduced your specimen to the required size and shape, the next thing to be done is to write a neat little label for it—the smaller the better—stating, first the nature of the specimen, secondly the geological formation to which it belongs, thirdly the locality from which it was procured, and fourthly the date when acquired, thus—

Limestone.
Lower Carboniferous.
Quarry, 1 mile N. W. of ――――
21. 8. 8-.

ruling a neat line at the top and bottom (red ink lines give a more finished appearance than black). When the label is dry, damp it to render it more pliant, and gum it on to the flattest available surface of the specimen, pressing it well into any small inequalities that it may hold the firmer. A small quantity of pure glycerine (about an eighth part) should be added to the gum before use, in order to prevent its drying hard and brittle. The specimen is now ready to place in its tray and be put away in the cabinet.

In the next place, pick out the fossils which you obtained from the limestone. With the cold chisel set in its block of wood, and the trimming hammer, remove as much of the surrounding rock (*matrix*) as you can without damaging the fossil, and with a smaller chisel any pieces that may be sticking to and obscuring it. Fossils in soft limestone, such as chalk, are best cleaned with an old penknife, and needles fixed into wooden handles, and finished off by the application of water with a nail-brush. Should you have the misfortune to break any specimen in the process of trimming, it should at once be mended. The most effectual cement for this purpose is made by simply dissolving isinglass in acetic acid, or, where the specimen contains much iron pyrites, and there would be a danger in starting decomposition, shellac dissolved in spirits of wine. When, however, neither of these are handy, chalk scraped with a penknife into a powder, and mixed with gum to the consistency of a thick paste, answers admirably. Failing this, however, gum alone will frequently suffice.

The next thing is to place the like kinds together in their several trays, writing a label, as before, for each tray, but leaving a blank space at the top for the insertion of the name when ascertained. The commoner sorts may be named from the figures of them given in the text-books (see list at the back of the title page); but failing this, it will be the best plan to seek the help of any friends who have collections, or to take the fossils to some museum, and compare them with the named specimens there exhibited. The label may be laid at the bottom of the tray with the fossils loose on the top of it, each fossil being marked with a number corresponding to one on the label.

Another plan is to fasten the label by one of its edges to the side of the tray ; or, if the fossils are small and mounted on a piece of card fitting into the tray, it may be gummed with them to the card.

Now let us take the shells we obtained from the dark-blue clay, with those and the bones from the old river bed up above. Gently turn them out of the tins, in which they were packed in the quarry, on to a paper or the lid of a card-board box, and with a pair of forceps pick them carefully out of the bran, and place them in large shallow trays, taking care not to mix those from the different beds. As we found when collecting them, these shells are extremely brittle from loss of animal matter, and our first object is therefore to harden them by some process, so that they will bear handling. To accomplish this you must get a saucepan, one of those wire contrivances for holding eggs when boiling, or a big wire spoon, such as formerly was used for cooking purposes, a packet of gelatine, and some flat pieces of tin, which last are easily procured by hammering out an old mustard or other tin, having previously melted in a gas flame the solder wherewith it is joined. Half fill the saucepan with clean water, and put in as much gelatine as when cold will make a stiff jelly ; melt this over the fire, placing the fossils meanwhile in a warm (not hot) corner of the fire-place ; then when the gelatine is quite dissolved, pile as many of them, whole or in pieces, into the egg-boiler, or spoon, as it will contain, hold them for a second in the steam, and then lower them gradually into the hot gelatine until it completely covers them. Little bubbles of air will rise and float on the surface. As soon as these cease to appear, raise the fossils above the surface and allow them to drip ; then pick them up one by one with the forceps, and spread them out on pieces of tin before the fire, but not too close to it. As soon as their exterior surfaces become dry, and before the gelatine gets hard, they should be taken up (they may be handled fearlessly now), and the superfluous gelatine sticking to the surface gently removed with a camel's-hair brush dipped in clean warm water ; otherwise, when dry, they present an unnatural varnished appearance, and have a tendency, on small provocation, to become unpleasantly sticky.

Small bones may be treated in like manner, but for large ones, weak glue is to be preferred to gelatine, which is only suitable for the finer and more delicate objects. Where it is desired to harden only a few things, it is better to mix the gelatine in a gallipot, which can be heated when required by standing it in a saucepan of water on the fire. In any case the gelatine need never be wasted, as it will keep almost any length of time, and can therefore be put by for future use. In default of the egg-boiler or wire-net spoon, an equally useful plan is to make a strainer from a piece of perforated zinc by turning up the edges all around, and attaching copper wire to it by which to lower the fossils into the gelatine, and raise them again.

When the fossils are quite dry they can be sorted, and those which have come to pieces may be mended with diamond cement (*i.e.* isinglass dissolved in acetic acid), and then properly labelled and placed in trays, or mounted as previously described.

To the plant remains and Lignite there is little that can be done beyond trimming them to suit the trays. Should there be much iron pyrites in the Lignite, it is sure, sooner or later, to decompose, when all that can be done is to throw it away. In the case, however, of valuable fruits and seeds, such as those from the London Clay of Sheppey, it is worth while to preserve them, if possible, in almost the only way known, viz. by keeping them in glycerine in wide-mouthed stoppered bottles, or by saturating them with paraffin.

Having prepared the specimens for the cabinet, the next thing is to arrange them in proper order. There are several ways of doing this, but for those who have not had much experience the following plan will be found the best :—Group the specimens according to the formations to which they belong, and arrange these groups in proper sequence (*vide* Table, p. 16); then take each group, and arrange the specimens it comprises in columns. Beginning at the top of the left-hand corner, place first the specimens of the rock itself, and under it any examples of minerals, concretions, etc., found in that rock ; next the fossil plants, if any ; and finally, such animal remains as you have arranged according to their zoological sequence, beginning with the lower forms (*vide* Table, p. 32). Unless cramped for room, each formation should begin a new box, its name being written on a slip of paper and placed at the head of the columns of trays. A label setting forth its contents should be fixed outside each of the boxes, which can then be put away on your cupboard shelves.

TABLE OF THE PRINCIPAL FOSSILIFEROUS STRATA ARRANGED IN CHRONOLOGICAL ORDER.

			Man.	Mammalia.	Birds.	Reptiles.	Amphibia.	Fishes.	Invertebrata.	Land Plants.
Quaternary, or Pleistocene.		Alluvial Deposits, River Valley Gravels and Cave Deposits. Drift and Glacial Deposits.								
Cainozoic, or Tertiary.		Pliocene. Miocene. Eocene.								
Mesozoic, or Secondary.	**Cretaceous.**	Chalk. Upper Greensand. Gault.								
	Neocomian.	Lower Greensand. Wealden.								
	Jurassic. — *Oolites.* Upper. Mid.	Purbeck. Portland. Kimmeridge Clay. Coral Rag. Oxford Clay.								
	Lower.	Cornbrash and Forest Marble. Great Oolite. Fullers' Earth. Inferior Oolite. Lias.								
Palæozoic, or Primary.	**Poikilitic.**	Trias, or New Red Sandstone. Permian.								
	Carboniferous.	Coal Measures. Millstone Grit and Yoredale Rocks. Carboniferous Limestone, etc.								
		Devonian and Old Red Sandstone.								
	Silurian.	Ludlow Beds. Wenlock Beds. Woolhope Beds. Tarannon Shale. Llandovery or May Hill Group.								
	Cambrian.	Bala and Caradoc Beds. Llandeilo Flags. Arenig Group. Tremadoc Slates. Lingula Flags. Menevian Beds. Longmynd and Harlech Group.								
		Pre-Cambrian and Laurentian.								

NOTES ON THE DIFFERENT FORMATIONS MEN-TIONED IN THE TABLE.

RECENT.—The alluvial deposits of most river valleys and some estuaries still in course of formation, containing fossil shells and mammals, all of living species.

QUATERNARY, POST-PLIOCENE, or PLEISTOCENE.—1. In-cluding the raised beaches around the coast, the older gravels of river valleys and the cave deposits, in all of which the shells are identical with those living in the rivers and seas of to-day, whilst the animals are many of them extinct, only a few being now found living on the spot.

2. The glacial drifts that cover all England north of the Thames, and which consist of sands, gravels, and clays, full of big angular stones frequently flattened on one side, scratched and sometimes polished from having been fixed in moving ice and forced over other rocks. A very interesting collection of these "boulders," as they are called, can be easily made, for they belong to almost every formation in England, and have some of them been brought from great distances, whilst the number and variety ob-tainable from a single pit is astonishing.

CAINOZOIC, or TERTIARY.—Beds of this age, in England at all events, are for the most part made up of comparatively soft rocks, gravels, sands, and clays, and are found in the eastern and south-eastern counties. They are divided into—

1. Pliocene, mainly consisting of a series of iron-stained sands, with abundant shell remains, and locally known as "crags." The shells are very partial in their distribution, the beds in places being almost entirely made up of them, whilst in others scarcely one is to be found. The great majority are of the same species as many still living. The Pliocene is subdivided into three groups:—

a. The *Norwich Crag Series*, sometimes called the "Mam-maliferous Crag," as at its base the bones of mastodon, elephant, hippopotamus, rhinoceros, and some deer have been found. The shells in it are such as still abound on the beaches of the eastern coast to-day—whelks, scallop shells, cockles, periwinkles, etc.

b. The *Red* or *Suffolk Crag*, its two names indicating its characteristic colour (a dark red-brown) and chief locality. From the base are obtained the celebrated phosphatic nodules miscalled "Coprolites," whence is manufactured an artificial manure, and with them are found the rolled and phosphatized bones and teeth of whales, sharks, etc. Amongst the shells the Reversed Whelks (*Fusus contrarius*), *Pecten opercularis*, *Pectunculus glycimeris*, several kinds of *Mactra* and *Cardium*, etc., are the commonest. Walton-on-the-Naze, Felixstowe, and Woodbridge are the best known localities.

c. The *White* or *Coralline Crag* is generally of a pale buff colour, and is in places almost entirely composed of the remains of Polyzoa. These (formerly called Corallines, whence the name Coralline Crag) are beautiful objects for a low-power microscope, or pocket lens, and are easily mounted in deep cells on slides. The bits of shell and sand that stick to them should be carefully removed with the point of a needle. A very large number of shells occur in this crag : of bivalves, the *Pecten* is very abundant, and its valves are frequently thickly grown over with Polyzoa ; *Cyprina Islandica*, *Cardita Senilis* are also plentiful ; and of univalves, the genus *Natica* is common. The Coralline Crag is best seen in the neighbourhood of Aldborough, Orford, Woodbridge, and other places in Suffolk.

2. Miocene, possibly represented in the British Isles by a small patch of clays and lignites at Bovey Tracey.

3. Eocene, divided into—

a. *Upper Eocene*, consisting of a series of very fossiliferous sands, clays, and limestones, exposed in the cliffs at the eastern and western ends of the Isle of Wight and on the neighbouring coast of Hampshire. They are partly of freshwater origin, when they contain the remains of freshwater shells such as *Limnæa Paludina*, *Planorbis*, etc. ; partly of marine origin, when shells belonging to such genera as *Ostrea*, *Venus*, *etc.*, take their place ; partly of estuarine, when the brackish water mollusca are found with bones and scutes of crocodiles and tortoises.

b. *Middle Eocene*, or the *Bagshot Beds*, composed of sands and clays. The beautiful coloured sands of Alum Bay, the sands of the Surrey and Hampstead Heaths, are familiar examples of the beds of this age. Very few fossils indeed have been found in them. The clay-beds on the contrary as seen at Barton and Hordwell on the Hampshire coast and again in the Isle of Wight, abound with shells belonging to genera such as *Conus*, *Voluta* and *Venus*, that inhabit warm seas. With them are the Nummulites, looking externally very like buttons, but on the inside divided into innumerable chambers in which the complex animal that formed the nummulite dwelt.

c. *Lower Eocene*, the well-known London clay, may almost be said to compose this division, for the underlying sands, gravels, and clays are in mass comparatively insignificant. The London clay contains plenty of fossils, only as they are disposed in layers (*zones*) at a considerable distance apart, they are not often hit upon. Layers of Septaria or cement-stones are of frequent occurrence. Sheppy is the great locality for London clay fossils, as the sea annually washes down large masses of the cliffs and breaks them up on the beach. A great many fossil fruits and seeds, remains of crabs, shells of Nautili, Volutes, and other mollusca, besides turtles, a species of snake, a bird with teeth, and a tapir-like animal, have at different times and in various places been found in this deposit, which sometimes attains a thickness of over 400 ft. The "Bognor Rock" is a local variety of the basement bed of this formation.

The MESOZOIC or SECONDARY rocks embrace a series of limestone, clays, sands, and sandstones that on the whole are well consolidated. The main mass of them lies to the west of a line drawn across the map of England from the mouth of the Tyne, in North-

Aturia Zic-zac (from the London clay).

umberland, southwards to Nottingham, and thence to the mouth of the Teign in Devonshire. In the south-eastern counties they underlie the tertiary rocks of the London and Hampshire basins, as they are called, at no great depth from the surface. Outlying patches of secondary rocks occur in Scotland, where they are found near Brora on the east coast, and in the islands of Skye and Mull on the west. In Ireland they are scantily represented round about the neighbourhood of Antrim. The secondary rocks are divided into—

1. Cretaceous.

a. The *Chalk* is too well known to need description, though technically it may be described as a soft white limestone chiefly built up of the microscopic shells of *Foraminifera*, and characterized in its upper part by nodules and bands of flint. These flints frequently inclose casts of fossils (sponges, sea-urchins, etc.), and sometimes shells themselves. Fossils, too, are fairly abundant, scattered throughout the mass. Amongst the commoner may be noticed the sea-

Ammonites various (from the chalk).

F

urchins, such as the "sugar loaf" (*Ananchytes*) and the heart-shaped *Micraster*, the Brachiopods or Lamp-shells (*Terebratula, Rhynchonella*), a "Thorny Oyster" (*Spondylus spinosus*), besides Ammonites, Belemnites (part of the internal shell of a kind of cuttle-fish), and the teeth of several species of sharks. Altogether the chalk is about 1,000 feet thick.

b. Upper Greensand is a series of greenish-grey sands and sandstones. The green colour, on close inspection, is seen to be due to the presence of innumerable small green grains of a mineral called glauconite. These are frequently casts of the chambers of the very same foraminifera that the chalk is so largely composed of.

Nodules and layers of "chert" (an impure kind of flint) occur in it, whilst in places it forms a hard rock called "firestone." The commonest fossils are Brachiopods, very similar to those in the chalk, a scallop-shell with four strongly marked ribs on it (*Pecten quadricostatus*), an oyster with a curved beak (*Exogyra columba*), and a pear-shaped sponge (*Siphonia pyriformis*). The Upper Greensand is better seen at places in the southern part of the Isle of Wight, in cliffs on the Dorsetshire coast, in Wiltshire, at Sidmouth, and in some parts of Surrey.

Rhynchonella depressa (a Brachiopod, from the Upper Greensand).

c. Gault, a stiff blue clay abounding in fossils: Ammonites often retaining their pearly shell; Belemnites, a bivalve with very deep furrows on it (*Inoceramus sulcatus*), and its first cousin (*I. concentricus*, p. 21), in which the ridge-like markings correspond with the lines of growth, besides many others, may be obtained in abundance from it. Layers of phosphatic nodules occur at irregular intervals. The gault is best studied at East Wear Bay, near Folkstone; it may also be seen in Dorsetshire, Wiltshire, and

Ammonites auritus (from the Gault).

Cambridgeshire; lately it has been found as far west as Exeter.

2. Neocomian.

a. The so-called *Lower Green Sand,* named in contradistinction to the *Upper Green Sand,* includes a series of iron stained sands, sandstones and clays of great thickness. The clayey beds are seen at Atherfield in the Isle of Wight, and at Nutfield in Surrey, while the sandy beds are met with at Speeton, at Folkestone, and near Reigate. Besides brachiopods and oysters, these beds have furnished a species of *Perna* (*P. Mulleti*), an elongated mussel (*Gervillia anceps*), a pretty *Trigonia* (*T. cordata*), some *Ammonites* and Nautili, with the teeth and bones of big reptiles. The cele-brated "Kentish Rag" and the sponge gravels of Farringdon are of this age.

b. Wealden. The main mass of these rocks occupies the area inclosed between the North and South Downs, and forms the Valley of the Weald,

Inoceramus concentricus (from the Gault).

whence they take their name. They consist of a series of sands, sandstones, clays, and shelly limestones that were deposited in the delta and off the mouth of a big river. The shells in them belong to freshwater genera, *Cyrena, Unio, Paludina,* etc. Bones of a huge lizard that hopped along on his hind legs (*Iguanodon*), and those of crocodiles, etc., are from time to time brought to light. The Wealden rocks occur also on both eastern and western sides of the Isle of Wight, and in Dorsetshire.

3. Oolites (or Roe-stones) are so named because the charac-teristic limestones of this formation resemble very much the roe of a fish. The small round grains, of which the typical examples are built up, when cut or broken through will be seen to be formed of numerous layers of carbonate of lime, disposed like the coats of an onion, around some central nucleus, generally a grain of sand, a fragment of coral, or the shell of one of the Foraminifera. They are divided into Upper, Middle, and Lower Oolites, and these again are subdivided as follows—

Upper Oolite.

a. Purbeck Beds, a series of fresh-water, with a few estuarine, or marine beds, which in point of fact connect the deposits we are next coming to with the Wealden just passed. **They** contain numerous fresh-water shells—*Paludina, Physa, Limnæa,* etc., with the microscopic valves of the little fresh-water crustacean *Cypris,* whose descendants are abundant in the rivers and lakes of to-day. An oyster occurs in the "Cinder Bed" and Plant remains in the "Dirt Beds." But the Purbecks are best known for the numerous remains of small mammals (*Plagiaulax*) allied to the kangaroo rat, at present living in Australia.

b. The *Portland Stone and Sand,* which come next in order, are largely quarried in the island whence they take their name. The quarrymen point out fossils in the stone, which they call "Horses'-heads" and "Portland screws." The former is the cast of a *Trigonia* shell; the latter, that of a tall spired univalve (*Cerithium*).

In Wiltshire, a coral (*Isastrea oblonga*) is found in the sandy beds, the original calcareous matter of which has been replaced by silex.

c. Kimmeridge Clay. This, by the pressure of the rocks subsequently deposited on it, has in greater part been hardened, and possesses a tendency to split in thin layers, and hence is termed by geologists a shale. It is seen at various points between Kimmeridge on the Dorsetshire coast and the Vale of Pickering in Yorkshire, and forms broad valleys. The principal fossils in it are Ammonites, a triangular-shaped oyster (*Ostrea deltoidea*), and one resembling a comma (*Exogyra virgula*).

Middle Oolites.

a. The *Coral Rag,* or *Coralline Oolite,* comprises a most variable set of beds, but principally a series of limestone, with fossil corals still in the position in which they grew, and resembling in form the reef-building corals of the Pacific. They rest on

b. Oxford Clay, a dark blue or slate-coloured clay without any corals, but containing a great many *Ammonites* and *Belemnites.* The *Kelloway Rock,* a sandy limestone at the base of the Oxford Clay, is well developed in Yorkshire, and furnishes amongst other fossils a large belemnite and an oyster (*Gryphæa dilatata*).

Lower Oolites.

a. Cornbrash, a very shelly deposit of pale-coloured earthy, and rubbly or sometimes compact limestone with plenty of fossils. The commonest are Brachiopods, Limas, oysters (*Ostrea Marshii*), Pholadomyas and Ammonites. It is best seen in Dorsetshire, Somersetshire, and near Scarborough in Yorkshire.

b. Forest Marble and *Bradford Clay.* The former is an exceedingly shelly limestone, often splitting into thin slabs. On the surfaces of some of the beds may be seen the ripple marks the sea made countless years ago, and the tracks of worms and crabs that dwelt in the mud or crawled on its surface at a time when it was soft mud. The Bradford clay is a very local deposit, taking its name from Bradford in Wiltshire, where it is most developed, and its characteristic fossil is the pear-shaped Encrinite or "stone-lily" (*Apiocrinus Parkinsoni*).

c. The *Great* or *Bath Oolite*, comprising a series of shelly limestones and fine Oolites, or freestones. The latter are largely quarried in the neighbourhood of Bath, and used for mantelpieces and the stone facings of windows. The great Oolite is rich in univalve mollusca, amongst which may be noted a limpet (*Patella rugosa*) and the handsome, tall-spired *Nerinæa Voltzii*, numerous bivalves belonging to the genera *Pholadomya Trigonia, Ostrea* (*O. gregaria*), and *Pecten*, besides Brachiopods (*Terebratula digona*, which looks very like a sack of flour, and *T. perovalis*, etc.).

At the base of the Great Oolite are the "Stonesfield slates," so-called—a series of thin shelly Oolites, etc., that split readily into very thin slabs. They are principally of interest to geologists on account of the discovery in them of the remains of small insect-feeding and possibly pouched mammals. With these are associated the bones of that big reptile the *Megalosaurus ;* the flying lizards called Pterodactyles ; fish teeth and spines ; lamp shells ; oysters, a *Trigonia* (*T. impressa*) ; and the impressions of insects, including a butterfly, and of plants.

d. Fullers' Earth, a clayey deposit occurring in the south-western parts of England, but not in the north. It abounds with a small oyster (*O. acuminata*) and Brachiopods (e.g. *Terebratula ornithocephala*), etc.

e. Inferior Oolite (including the Midford Sands). As these beds are followed across the country from the south-west of England to Yorkshire, they are found to change greatly in character. Limestone and marine beds in the south are replaced by sandy and estuarine beds in the north. Amongst other fossils from beds of this age may be found several Echinoderms, a crinkly lamp shell (*Terebratula frimbriata*), and a spiny one (*Rhynchonella spinosa*), bivalves belonging to the Genera *Ostrea, Trigonia, Pholadomya*, etc., and some very handsome Ammonites (e.g. *A Humphre-sianus*).

4. Lias.
This for the most part consists of very regular alternations of argillaceous (clayey) limestone and clay, or shale. It is of great thickness, and hence for convenience has been divided into (a) *Upper Lias*, (b) *Middle Lias* or *Marl-stone*, and (c) *Lower Lias*. A large number of fossils are to be found in it.

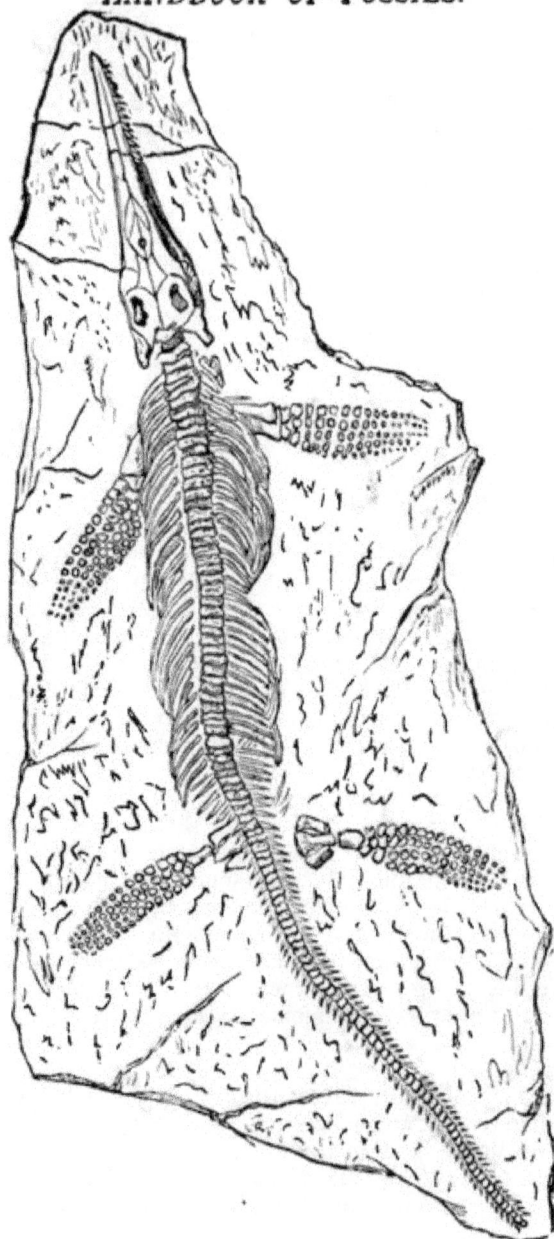

Ichthyosaurus, or Fish-lizard (from the Lias).

Lyme Regis and Whitby are perhaps the best known localities ;

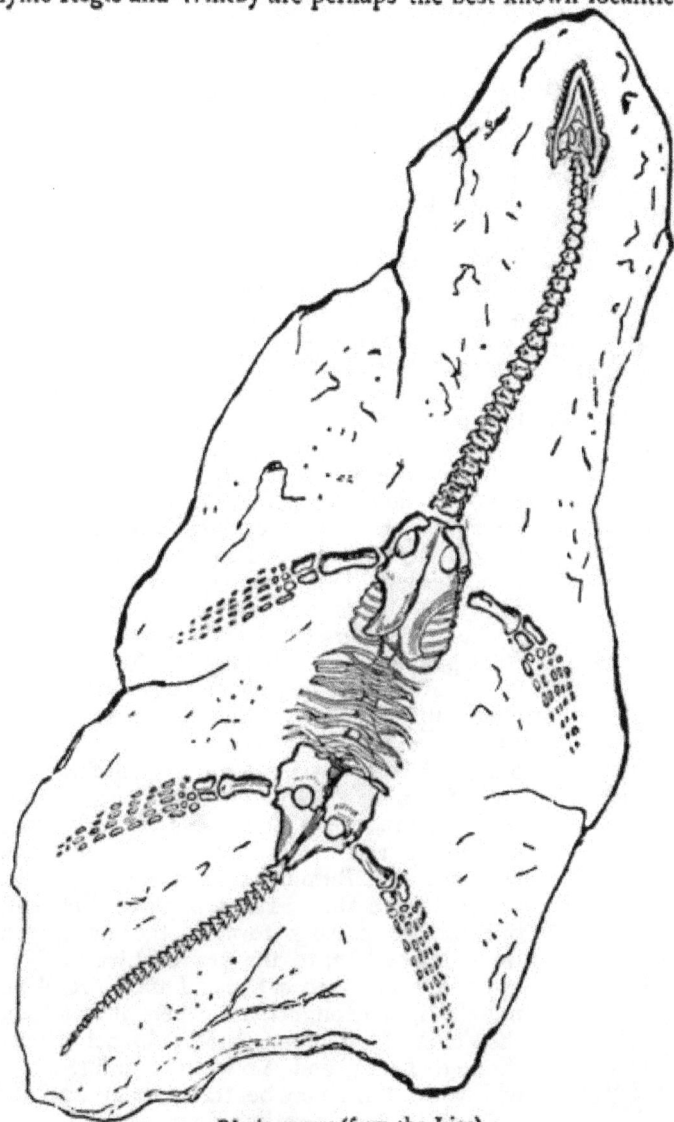

Plesiosaurus (from the Lias).

the former, on account of the great number of specimens obtained of the huge fish-lizard (*Ichthyosaurus*, p. 24), and long-necked *Plesiosaurus* (p. 25), besides numberless fish; whilst the latter is renowned for its jet (or fossilized wood) and its "snake-stones" (*Ammonites*), concerning which curious old stories are told. *Ammonites* are plentiful in the Lias, which has been sub-divided into zones, or layers, named after the ammonite occurring in greatest numbers in that particular zone. There is one thin limestone band in the Marlstone composed entirely of the shells of *Ammonites planicostatus*. A curious kind of oyster (*Gryphæa incurva*), locally known as the devil's toenail, a huge *Lima* (*L. gigantea*), a magnificent Encrinite (*Extracrinus Briareus*), and numerous other fossils, are also to be obtained by patient search.

5. Rhætic, Penarth Beds, or White Lias. These beds are not of any considerable thickness, but are very persistent, and of great interest, inasmuch as they yield the remains of the oldest known mammal (*Microlestes*), a small insect-feeder. They are composed of limestones, shales and marls (*i.e.* limey clays), and are best studied in Somersetshire and Dorsetshire. The "landscape marble" belongs to this formation, which also contains a bone bed, or thin layer made up of the bones and teeth, etc., of fish. Shells are not numerous, though the casts of one species (*Avicula contorta*) is plentiful.

6. Trias, or New Red Sandstone, a thick series of sandstones and marls, the great mass of which forms the subsoil of the western midland counties, Birmingham being nearly in the centre, thence they extend in three directions, one branch passing towards the north-west, through Cheshire, to the sea at Liverpool, re-appearing on the coast line of Lancashire, Westmoreland, and Cumberland, where it also forms the Valley of the Eden. Another branch extends through Derby and York to South Shields, whilst the third may be traced southwards in isolated patches down into Devonshire.

Belemnitas elongatus (from the Lias).

There are scarcely any fossils in it, but in Worcestershire and Warwickshire the bivalve shell of a small crustacean (*Estheria minuta*) occurs in the upper beds; whilst now and again the teeth and bones of some strange amphibians (*Labyrinthodon*), or the impressions of their feet (*Cheirotherium*) where they crawled on the then soft mud of the foreshore, are found. The Trias is divided into Upper Trias or Keuper, and Lower Trias or Bunter. The middle beds (Muschelkalk), which are found in Germany, where they contain plenty of fossils, are wanting in this country. In the lower beds of the Keuper, layers of rock salt, sometimes of great thickness, occur, whilst casts (called pseudomorphs) of detached salt-crystals are found abundantly in the sandy marls. Northwich, Nantwich, Droitwich, and several other towns in Cheshire and

Ceratites nodosus
(from the Muschelkalk).

Worcestershire, are famed for their salt works, the salt being either mined or pumped up as brine from these beds.

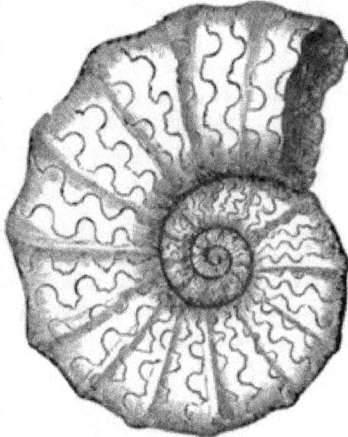

PALÆOZOIC OR PRIMARY.—Beds of this age generally possess a more crystalline and slaty structure than any of those already mentioned, are usually more highly inclined and disturbed, and form for the most part more elevated ground. They are the principal store-houses of our mineral wealth, containing as they do coal, iron, and other metals. The Palæozoic rocks are found in England to the north and west of the secondary series, beneath which they disappear when traced to the south-east. Wales, and the greater part of Scotland and Ireland, consist of beds of this age.

1. Permian. Under this term are included beds of red sandstones and marls, closely resembling those of Trias, and like them containing but few fossils, as well as a very fossiliferous limestone, known as the Magnesian Limestone, from the abundance of magnesia it contains. A pretty polyzoan (*Fenestella retiformis*), a spiny brachiopod (*Productus horridus*), various genera of fish, chiefly found in a marl state underlying the limestone, some Labyrinthodonts and plant remains, are the principal forms met with in this formation.

2. Carboniferous. This, from a commercial point of view, is

the most important of all the formations, comprising as it does the coal-bearing strata. It is subdivided into—

a. Coalmeasures, a series of sandstones and shales with which are interstratified the seams of coal, varying in thickness from six inches to as much in one instance as thirty feet.

Coal is the carbonized remains of innumerable plants, chiefly ferns and gigantic clubmosses, that grew in swamps bordering on the sea-coast of the period. Each coal seam is underlain by a bed of clay called "under-clay," containing the roots of the plants that grew on it. Some of the best impressions of ferns, etc., are to be obtained in the shaley beds forming the roof of the coal seam ; many good specimens, however, are to be got by searching the refuse heap at the pit's mouth. Besides plants, the remains of fish are abundant in some of the beds of shale. And in Nova Scotia the bones of air-breathing reptiles and land snails have been discovered. Cockroaches and other insects were also denizens of the carboniferous forests.

The following are the principal coalfields :—
1. Northumberland and Durham coalfield.
2. South Lancashire coalfield.
3. Derbyshire coalfield.
4. Leicestershire and Staffordshire coalfields.
5. South Wales coalfield.
6. Bristol and Somerset coalfields.

b. Millstone grit or *Farewell-rock.* The former term explains itself, the latter designation has been applied to it in the southern districts, because when it is reached, then good-bye to all work-able coal-seams.

It consists of coarse sandstones, shales, and conglomerates with a few small seams of coal. Fossils are not very common in it.

c. Yoredale Rocks, a series of flagstones, gritstones, limestones and shales, with seams of coal, occurring in the northern counties. It is underlain by—

d. Carboniferous or *Mountain Limestone,* which in places is upwards of 1,000 feet thick, and full of fossils. The stems of encrinites, or "stone-lilies," corals, brachiopods (*e.g. Productus, Orthis,* etc.), and Mollusca, including some Cephalopods, like *Goniatites* and the straight Nautilus (*Orthoceras*), with fish teeth, etc., go to compose this tough, bluish-grey limestone which is largely quarried for marble mantlepieces, etc.

e. The *Tuedian group* in the north, and *Lower Limestone Shale* in the south, follow next, and consist of shales, sandstones, limestones, and conglomerates, varying greatly in different districts, and containing few fossils.

3. Devonian or Old Red Sandstone. To this age are assigned a perplexing series of strata, the principal members of which consist of (*a*) a thick limestone, well seen in the cliffs and marble quarries of south Devon, and full of fossil-corals (*e.g. Favosites polymorpha* [or *cervicornis*]) Brachiopods, and Mollusca, etc.

b. A series of sandstones, slates, and limestones in North Devon containing Trilobites (*Phacops, Bronteus,* etc.), Brachiopods, and other fossils.

c. The *Old Red Sandstone* of Wales, the North of England, and Scotland, consisting of red and grey sandstone and marly beds, with remains of fish.

These fish, unlike most now living, were more or less covered with hard external plates, and possessed merely a cartilaginous skeleton. In one set of individuals, indeed (*Pterichthys*), the armour plates formed quite a little box. These creatures propelled themselves by means of two arm-like flippers, rather than fins. They were but a few inches long, and appear pigmies in contrast to the strange half-lobster-like crustacean, *Pterygotus,* that lived with them, and attained sometimes as much as five feet in length.

4. Silurian. Named by Sir Roderick Murchison after a tribe of Ancient Britons that dwelt in that part of Wales, where these rocks were first observed. Some of Murchison's Lower Silurian beds were included by Professor Sedgwick in his Cambrian, of which we shall have to speak next ; and as these two geologists never could agree on a divisional line between their respective formations, and since succeeding observers have followed sometimes one and sometimes the other method of classification, considerable confusion has resulted. Here, however, for several reasons, we propose to follow Sedgwick's arrangement ; and hence, under the term Silurian, retain only Murchison's Upper beds. They consist of a series of sandstones, gritstones, conglomerates, shales, limestones, etc.

Amongst the more important fossils, which are very abundant in the limestones, are various corals (*e.g.* the Chain-coral *Halysites*), Star-fish, Crinoids, Trilobites (*Phacops,* etc.), Polyzoa, Brachiopods and Mollusca, especially Cephalopoda (*Orthoceras, Nautilus,* etc.).

These rocks occur principally in the border land between England and Wales, and the adjacent counties ; but are also

represented in Westmoreland, Scotland, and Ireland. Their
principal subdivisions are given in the Table on p. 16.

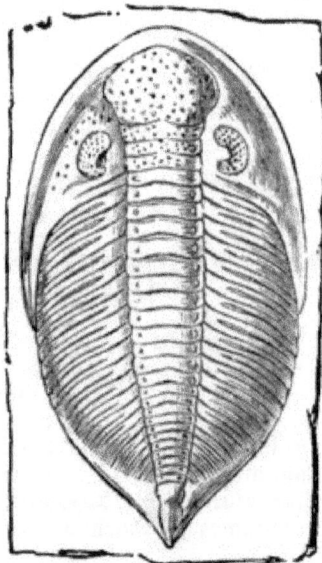

TRILOBITE (*Asaphus caudatus*),
(from the Silurian).

Orthoceras subannulatum (from the Silurian).

5. Cambrian. Under this term, derived from the old name
for Wales, are included many sandstones, grits, slates and flags,
with here and there a limestone band. They form the greater
part of the western counties of Wales, where they rise to a considerable height above the sea level. The highest hills of
Westmoreland and more than half of Scotland are composed of
beds of this age.

The fossils, save in the limestone bands, are not easy to find,
but in places they are fairly abundant. Brachiopods are far
more numerous than the Mollusca properly so-called. Of these,
the genus *Orthis* was most abundant at about the close of this
period. Certain beds of this age have received the name of
Lingula Flags, owing this prevalence in them of the curious
Brachiopod *Lingula* so like the species now living in some of the
warm seas of the tropics. The Trilobites included several forms,
and one species (*Paradoxides Davidis*) attained the length of
nearly two feet. A few star-fish, some Hydrozoans (*Graptolites*),
and the tubes and casts of Annelides and tracks of Trilobites,

complete the list of more remarkable fossils. The subdivisions of the Cambrian rocks will be found in the table on p. 16.

6. Pre-Cambrian.—Near St. David's Head and some other places in Wales, in Anglesea, Shropshire, etc., some yet older rocks have been found. They are probably for the most part of volcanic origin, but they have been so much changed since they were first deposited, and as hitherto no fossils have been found in them, little is known concerning them.

Parts of the western coast of Northern Scotland and the Hebrides are composed of a crystalline rock called Gneiss, and supposed to be the oldest member of the British strata. No fossils have been found in it.

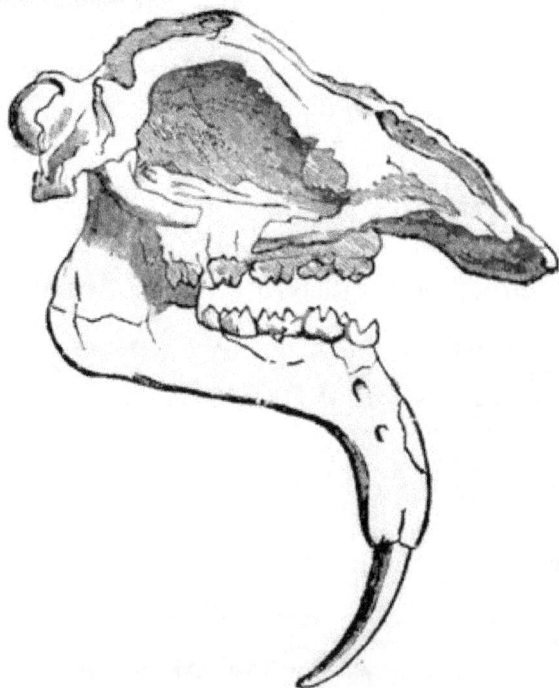

Skull of *Deinotherium giganteum*, a huge extinct animal, related to the elephants (from the Miocene of Germany).

VOLCANIC ROCKS. Although there are fortunately no volcanoes to disturb the peace of our country at the present day, there is abundant evidence of their existence in the past. Not only are some of the beds, especially those of Paleozoic age, composed of the dust and ashes thrown out of volcanoes, with here and there a lava flow now hardened into solid rock, but the stumps of the volcanoes themselves are left to tell the tale. The cones indeed are gone, carried off piecemeal by the rain and frosts, and other destructive agencies, in the course of countless ages : not so the once fluid rock within ; *that* cooled down into Granite, and though originally below the surface, it now, owing to the removal of the overlying softer strata, forms raised ground overlooking the surrounding country. The granite masses of Cornwall, of Dartmoor, in the south-west of Mt. Sorrel ; the variety called Syenite at Malvern and Charnwood Forest ; the Basalts of the Cheviot Hills and of Antrim ; the volcanic rocks of Arthur's Seat, Edinburgh, and of the islands of Skye and Mull, etc., are examples of this class of rock. They are of different ages, and belong to different periods of the earth's history, from early Palæozoic down to Miocene times.

TABLE OF THE PRINCIPAL DIVISIONS OF THE ANIMAL KINGDOM, TO SHOW THE ORDER IN WHICH THE FOSSILS SHOULD BE ARRANGED.

INVERTEBRATA.

Foraminifera, minute chambered shells like the Nummulite.
Spongida, Sponges.
Hydrozoa, Graptolites, etc.
Actinozoa, Corals.
Echinodermata, Sea-urchins, Stone-lilies, Starfish, etc.
Annelida, Worm tracks.
Crustacea, Trilobites, Crabs, etc.
Arachnida, Scorpions and Spiders.
Myriapoda, Centipedes.
Insecta, Beetles, Butterflies, etc.
Polyzoa (*Bryozoa*) or Moss Animals.
Brachiopods, Lampshells.

Mollusca
{ *Lamellibranchiata*, Bivalves.
{ *Gasteropoda*, Univalves.
{ *Cephalopoda*, Cuttlefish, Ammonites.

VERTEBRATA.

Pisces, Fish.
Amphibia, Labyrinthodonts, Frogs, and Newts.
Reptilia, Reptiles.
Aves, Birds.
Mammalia, Mammals.

WORKS OF REFERENCE.

FOR NAMING COMMON FOSSILS.

Tabular View of Characteristic British Fossils Stratigraphically Arranged. By J. W. LOWRY. *Soc. Prom. Christ. Knowledge.* 1853.

Figures of the Characteristic British Tertiary Fossils (Chiefly Mollusca) Stratigraphically Arranged. By J. W. LOWRY and others. *London (Stanford).* 1866.

PALÆONTOLOGY.

The Ancient Life History of the Earth. By H. A. NICHOLSON. 8vo. *Edinburgh and London.* 1877.

A Manual of Palæontology. By H. A. NICHOLSON. 2nd edition. 2 vols. 8vo. *Edinburgh and London.* 1879.

PETROLOGY.

The Study of Rocks. By F. RUTLEY. (Text Books of Science.) 8vo. *London.* 1879.

FIELD GEOLOGY.

A Text-Book of Field Geology. By W. H. PENNING. With a Section on Palæontology, by A. J. JUKES-BROWN. 2nd edition. 8vo. *London.* 1879.

GEOLOGY IN GENERAL.

The Student's Elements of Geology. By SIR CHARLES LYELL, Bart. 4th edition. 8vo. *London.* 1884.

The Principles of Geology. By SIR CHARLES LYELL, Bart. 12th edition. 2 vols. 8vo. *London.* 1875.

Phillip's Manual of Geology. 2nd edition. By SEELEY AND ETHERIDGE. 2 vols., 8vo. *London.* 1885.

Tabular View of Geological Systems, with their Lithological Composition and Palæontological Remains. By D. E. CLEMENT. *London (Sonnenschein).* 1882.

BRITISH GEOLOGY.

The Physical Geology and Geography of Great Britain. By SIR ANDREW C. RAMSEY. 5th edition. 8vo. *London.* 1878.

The Geology of England and Wales. By HORACE B. WOODWARD. 8vo. *London.* 1876.

Geology of the Counties of England and Wales. By W. J. HARRISON. 8vo. *London.* 1882.

POPULAR
ILLUSTRATED SCIENTIFIC BOOKS,

PUBLISHED BY

SWAN SONNENSCHEIN & CO.

UNIFORM WITH THIS VOLUME.

ALL FULLY ILLUSTRATED.

BRITISH BUTTERFLIES, MOTHS, AND BEETLES.
By W. F. KIRBY (Brit. Mus.). Crown 8vo, cloth, 1s.

MOSSES, LICHENS, AND FUNGI.
By PETER GRAY and E. M. HOLMES. Crown 8vo, cloth, 1s.

ENGLISH COINS AND TOKENS.
By LLEWELLYNN JEWITT, F.S.A.; with a chapter on **Greek and Roman Coins**, by BARCLAY V. HEAD, M.R.A.S. Crown 8vo, cloth, 1s.

FLOWERS AND FLOWER LORE.
By Rev. HILDERIC FRIEND, F.L.S. Illustrated. Third Edition, demy 8vo, cloth gilt, 7s. 6d.

THE DYNAMO: How Made and How Used.
By S. R. BOTTONE. Numerous Cuts. Crown 8vo, cloth, 2s. 6d.

A SEASON AMONG THE WILD FLOWERS.
By Rev. H. WOOD. Illustrated. Crown 8vo, cloth gilt, 2s. 6d.

HISTORY OF BRITISH FERNS.
By E. NEWMAN, F.L.S. Fifth Edition, Illustrated. 12mo, cloth, 2s.

THE INSECT HUNTER'S COMPANION.
By Rev. J. GREENE. Third Edition. Cuts. 12mo, boards, 1s.

TABULAR VIEW OF GEOLOGICAL SYSTEMS.
By Dr. E. CLEMENT. Crown 8vo, limp cloth, 1s.

SWAN SONNENSCHEIN & CO., PATERNOSTER SQUARE.

Paternoster Square.]　　　　　　[*April 1st*, 1886.

𝖲𝖾𝗅𝖾𝖼𝗍𝖾𝖽 𝖫𝗂𝗌𝗍 of
SWAN SONNENSCHEIN & Co.'s
𝔜𝔲𝔟𝔩𝔦𝔠𝔞𝔱𝔦𝔬𝔫𝔰.

For **SUBJECT INDEX,** *see pp.* 27—32.

Abdy-Williams. Novels by E. M. ABDY-WILLIAMS :

Two Ifs. 1 vol. ed. Cr. 8vo. 3*s*.6*d*.	For his Friend. 3 vols. Cr. 8vo
Forewarned ! Fcap. 8vo. 1*s*.	31*s*. 6*d*.

Adams. Novels by MRS. LEITH-ADAMS. Cheap Editions.

Each vol. cr. 8vo, *cloth extra*, 3*s*. 6*d*.

Geoffrey Stirling.	Cosmo Gordon.	[*Shortly*.
Madelon Lemoine.	Lady Deane.	[*Shortly*.

Adams, Rev. F. A. My Man and I; or, the
Modern Nehemiah. 8vo, *cloth*, 7*s*. 6*d*.

Adams. Books by W. H. DAVENPORT ADAMS:

A Book of Earnest Lives.
With 8 portraits and plates.
Demy 8vo, *cloth gilt, gilt edges,*
7s. 6d.
    Dean Colet, Roger Ascham,
    Lady Mary Montagu, Robert
    Raikes, Lord Brougham, Dr.
    Arnold, J. F. Oberlin, Mary
    Carpenter, Wm. Wilberforce,
    Sir T. F. Buxton, John
    Eliot, John Howard, Mrs.
    Fry, Mrs. Mompesson, Sister
    Dora, and others.
Battle Stories from English
and European History. *Second
Edition.* With 16 plates and
plans. Demy 8vo, *cloth gilt, gilt
edges,* 7s. 6d.
    European: Byzantium, Cressy,
    Poictiers, Navarette, Agin-
    court, Lützen, Blenheim,
    Malplaquet, Pultowa, Water-
    loo, Inkerman. *English:*
    Hastings, Falkirk, Bannock-
    burn, Bosworth Field, Flod-
    den Field, Marston Moor,
    Naseby, Culloden. *Anglo-
    Indian:* Plassey, Haiderabad,
    Guzerat; and others.

**Girlhood of Remarkable
Women.** *Second Edition, enlarged.* With 8 plates. Demy
8vo, *cloth gilt, gilt edges,* 7s. 6d.
    Harriet Martineau, Fanny
    Burney, Elizabeth Inchbald,
    Charlotte Brontë, Sara Coler-
    idge, Mary Somerville, Mary
    Russell Mitford, Lady Mor-
    gan, Lady Jane Grey, Mrs.
    Hutchinson, Countess of
    Pembroke, Margaret More,
    Lady Mary Montagu, Cath-
    erine of Siena, Jeanne d'Arc,
    Mme. de Miramon, Eliz.
    Carter, Caroline Herschel,
    Lady Fanshawe, and others.
Celebrated Women Travellers
of the Nineteenth Century.
Second Edition. With 8 plates.
Cr. 8vo, *cloth gilt, gilt edges,*
3s. 6d.
    Frederika Bremer, Ida Pfeiffer,
    Lady Stanhope, Lady Brassey,
    Lady Morgan, Mrs. Trollope,
    Isabella Bird, Lady Florence
    Dixie, Miss Gordon Cumming,
    Lady Barker, and others.

Alberg. Books by ALBERT ALBERG:

Gustavus Vasa and his Stir-
ring Times. *Third Edition.*
Illustrated. Cr. 8vo, *cloth gilt,
gilt edges,* 3s. 6d.

Charles XII. and his Stirring
Times. Illustrated. Cr. 8vo,
cloth gilt, gilt edges, 1s. 6d.

"Alert." Cruise of H.M.S. "Alert." Four Years in
Patagonian, Polynesian, and Mascarene Waters. By R. W.
Coppinger, M.D. (Staff-surgeon on board). With 16 Plates and
several cuts in the text from drawings and photos. by the Author
and F. North, R.N. *Fourth Edition.* 8vo, *cloth, gilt edges,* 6s.

Allen, Grant. The Evolution of Flowers.
[*In preparation.*

Alpine Plants. *See* Seboth and Bennett; and Bennett.

Althaus. *See* Schaible and Althaus.

Andersen, Hans. Fairy Tales Set to Music by Annie Armstrong. 4to, *cloth,* 1s. 6d. ; *paper,* 1s.

Arabian Nights, The New. Select Tales, not included in the editions of Galland or of Lane. Translated by W. F. Kirby (British Museum). *Second Edition.* Illustrated. Cr. 8vo, *cloth gilt, gilt edges,* 3s. 6d.

Armstrong, J. Birds and their Ways. Illustrated. Cr. 8vo, *cloth gilt, gilt edges.* 1s. 6d.

Arthur, T. S. Ten Nights in a Bar-Room. *New Illustrated Edition.* Cr. 8vo, *cloth gilt,* 2s.

Auerbach, Berthold. Two Stories (Christian Gellert and The Stepmother). Cuts. Cr. 8vo, *cloth gilt, gilt edges,* 2s. 6d.

Austin, Jane G. Moonfolk : A True Account of the Home of the Fairies. *Second Edition.* Illustrated by W. J. Linton. Cr. 8vo, *cloth gilt, gilt edges,* 2s. 6d. [*Fairy Library.*

Axon, W. E. A. Gipsy Folk Tales. [*In preparation.*

Babcock, W. H. Cypress Beach : A Novel. 2 vols. Cr. 8vo, *cloth,* 12s.

Bagnall, J. E. Handbook of Mosses. With Numerous Woodcuts. Cr. 8vo, *cloth,* 1s.

Baker, Ella. Stories of Olden Times. Drawn from History and Tradition. *Second Edition.* Cr. 8vo, *cloth gilt, gilt edges,* 1s. 6d.

Barras. Works by COLONEL JULIUS BARRAS.

India, and Tiger Hunting. 2 vols. cr. 8vo, ea. 3s. 6d.	The New Shikari at our Indian Stations. 2 vols. cr. 8vo, ea. 3s. 6d.

Baxter, Right Hon. W. E., M.P. England and Russia in Asia. Cr. 8vo, *cloth,* 1s. [*Imp. Parl. Ser.*

Bennett, A. W., M.A., B.Sc. Tourists' Guide to the Flora of the Alps. Edited from the work of Prof. K. W. v. Dalla-Torre, and issued under the auspices of the German and Austrian Alpine Club in Vienna. Elegantly printed on very thin but opaque paper, 392 pp., bound as a Morocco pocket-book. Pocket size, 5s.

Best Books, The. A Classified Bibliography of the Best Curren English and American Literature, with the Publishers' names, the prices, and the dates of each work. 4to. [*Shortly.*

Bevan, J. A., M.D. The March of the Strikers. 1s.

Bevan. Works by G. PHILLIPS BEVAN, F.G.S., F.S.S. :

The Royal Relief Atlas of all parts of the Globe, consisting of 31 Embossed Maps. *Second Edition.* Royal 4to, 21*s.* Each Map is separately framed, and the whole bound in one volume, *half persian.*

The Home Geography. [*In preparation.*
Guide to Lichfield Cathedral. [*In preparation.*
Guide to Westminster Abbey. [*In preparation.*

Bickerdyke. Works by John BICKERDYKE, M.A. :

With the Best Intentions: A Tale of Undergraduate Life at Cambridge. Cr. 8vo, *cloth,* 2*s. 6d.*

An Irish Midsummer Night's Dream. Frontispiece. Cr. 8vo, *cloth gilt, gilt edges,* 1*s. 6d.*

Birthday Book, The Floral. By FLORENCE DUDGEON. With Coloured Plates. Imp. 16mo, *cloth gilt, gilt edges,* 3*s. 6d.*

Boger, C. G. Elfrica : An Historical Romance of the Twelfth Century. 3 vols. Cr. 8vo, *cloth,* 31*s. 6d.*

Bottone, S. R. The Dynamo : How Made and How Used. Numerous Cuts. Cr. 8vo, *cloth,* 2*s. 6d.*

Bourne. Works by C. E. BOURNE, Barrister-at-law :

Heroes of African Discovery and Adventure.
  Series I.—To the Death of Livingstone.
  Series II.—To 1883.
  Each series in a *Second Edition.* With Plates and Coloured Maps. Cr. 8vo, *cloth gilt, gilt edges,* each 3*s. 6d.*
The Great Composers : Short Lives of Eminent Musicians.

Second Edition. With portraits and plates. Cr. 8vo, *cloth, gilt edges,* 3*s. 6d.*
  Händel — Bach — Gluck — Haydn — Mozart—Beethoven —Weber—Schubert—Rossini — Mendelssohn — Chopin — Schumann—Berlioz.
Life of Gustavus Adolphus. Illustrated. Cr. 8vo, *cloth gilt, gilt edges,* 1*s. 6d.*

Bowker, J. Goblin Tales of Lancashire. Illustrated by Charles Gliddon. Cr. 8vo, *cloth gilt, gilt edges,* 2*s. 6d.*
[*Fairy Library.*

Bradshaw. Works by Mrs. JOHN BRADSHAW.

Roger North. 3 vols. Cr. 8vo, 31*s. 6d.*

Merevale : A Novel. 1 vol. 6*s.*

Brant, Elizabeth, Head Mistress of the Granby Schools. Systematic Cutting-out for the New Code, from Units of Measurement. With Folding Diagrams in red and black. Cr. 8vo, *cloth,* 2*s.*

Broadhurst and Reid. Leasehold Enfranchisement. By HENRY BROADHURST, M.P., and R. T. REID, M.P. Cr. 8vo, *cloth,* 1*s.* [*Imp. Parl. Ser.*

Bulow. Works by the Baroness BÜLOW:

The Child and Child-Nature. *Third Edition.* Cuts. Cr. 8vo, *cloth*, 3*s*.
[*Kindergarten Manuals.*

Hand-Work and Head-Work: Their Relation to One Another. Cr. 8vo, *cloth*, 3*s*.
[*Kindergarten Manuals.*

Burke, Ulick J. Couleur de Rose: A Novel. 2 vols. Cr. 8vo, *cloth*, 21*s*.

Butler, E. A. The Entomology of a Pond.
[*In preparation.*

Buxton, Sydney, M.P. Over-Pressure and Elementary Education. Crown 8vo, 2*s*. ; paper, 1*s*.
See also Imperial Parliament Series, p. 27.

Caballero, Fernan. Book of Spanish Tales. *Third Edition.* Illustrated by Chas. Harrison. Cr. 8vo, *cloth gilt, gilt edges*, 2*s*. 6*d*.
[*Fairy Library.*

Caine, Hoyle, and Burns. Local Option. By W. S. CAINE, M.P., WM. HOYLE, and Rev. DAWSON BURNS, D.D. *Second Edition.* Cr. 8vo, *cloth*, 1*s*.
[*Imp. Parl. Ser.*

Cambridge Examiner, The. A Monthly Educational Journal (except in July and August). Demy 8vo, Vols. I.—V. [1885], each 5*s*.
[48 pages, *Monthly*, 6*d*.

Camden. Tales by CHARLES CAMDEN:

The Travelling Menagerie. Illustrated by J. Mahoney. Sm. 8vo, *cloth gilt, gilt edges*, 2*s*. 6*d*.

Hoity Toity, the Good Little Fellow. Illustrated by J. Pettie, R.A. Sm. 8vo, *cloth gilt, gilt edges*, 2*s*. 6*d*.

Cappel, E. S. Old Norse Sagas. Illustrated. Cr. 8vo, *cloth gilt, gilt edges*, 2*s*. 6*d*.
Aslog—Frithiof—Ingeborg—Ragnar Lodbrok—Sigurd—Wayland Smith—Hamlet—and others.
[*Fairy Library*

Chapman. Books by WILLIAM CHAPMAN:

Notable Women of the Covenant. With portraits and plates, Cr. 8vo, *cloth gilt, gilt edges,* 2*s*. 6*d*.
Notable Women of the Puritan Times. With portraits and plates. Cr. 8vo, *cloth gilt, gilt edges*, 3*s*. 6*d*.

Notable Women of the Reformation. With portraits and plates. Cr. 8vo, *cloth gilt, gilt edges*, 3*s*. 6*d*.
Life of Martin Luther. Cuts Cr. 8vo, *cloth gilt, gilt edges,* 1*s*. 6*d*,
Life of John Wiclif. Cuts. Cr. 8vo, *cloth gilt, gilt edges,* 1*s*. 6*d*.

Children's Journey, The, and other Stories. By the Author of "Voyage en Zigzac," etc. *Second Edition.* Illustrated by the Author. Cr. 8vo, *cloth gilt, gilt edges,* 3s.

Chitty. Coloured Books by LILY CHITTY :

Harlequin Eggs. Twenty-four coloured plates by Lily Chitty. With text by Ismay Thorn. 4to, 2s. 6d.

In and Out. Twenty-four coloured plates by Lily Chitty. With text by Ismay Thorn. 4to, 2s. 6d.

Chitty, W. Practical Beekeeping. 12mo. *At press.*

Churchill, Mrs. Spencer. Daisy Darling. A Novel. Cr. 8vo, *cloth,* 3s. 6d.

Church Rambles and Scrambles. Cr. 8vo, *cloth,* 2s.

Clarke. Short Biographies by F. L. CLARKE. Each vol. illustrated. Cr. 8vo, *cloth gilt, gilt edges,* 1s. 6d.

A Book of Golden Friendships. Illustrated. Cr. 8vo, *cloth gilt, gilt edges,* 3s. 6d.
Childhood of the Prince Consort.
Charlemagne and his Times
Life of William Tyndale.

George Stephenson.
Robert Stephenson.
Lives of George and Robert Stephenson (in 1 vol). *Cloth gilt, gilt edges,* 2s. 6d.
Sir Walter Raleigh and his Times.

Claus — Sedgwick. Elementary Text Book of Zoology. By Prof. W. CLAUS, edited by ADAM SEDGWICK, M.A., Fellow and Lecturer of Trin. Coll., Cambridge, assisted by F. G. Heathcote, B.A., Trin. Coll., Camb. Illustrated by 706 woodcuts drawn by Prof. Claus. In Two Parts. Demy 8vo, *cloth.*

Part I. Protozoa to Insecta. 21s. | Part II. Mollusca to Man. 16s.

Clement, Dr. E. Tabular View of Geological Systems. Cr. 8vo, *limp cloth,* 1s.

Cobbe, Lucy. Doll Stories. Cuts. *Cloth, gilt edges,* 1s. 6d.

Cockburn, Dr. Samuel. The Laws of Nature and the Laws of God : a Reply to Prof. Drummond. Cr. 8vo, *cloth,* 3s. 6d.

Conn, William. From Paris to Pekin over Siberian Snows. Edited from the Travels of Victor Meignan by William Conn. With 16 plates. Demy 8vo, *cloth extra,* 16s.

Contemporary Pulpit, The. A Monthly Homiletic Magazine. Vols. I.—IV. Royal 16mo, *cloth extra, gilt top, each* 6s.
[64 pages, *Monthly,* 6d.

Cooke. Short Biographies by FRANCES E. COOKE. Each vol. illustrated. Cr. 8vo, *cloth gilt, gilt edges.*

A Boy's Ideal. [Life of Sir Thomas More.] 1s. 6d.

True To Himself. [Life of Savonarola.] 1s. 6d.

Latimer's Candle. [Life of Latimer.] 1s. 6d.

An English Hero. [Life of Richard Cobden.] 1s. 6d.

Cooper, A. J. **Training of the Memory.** 12mo. 3d.

Corbett, Mrs. George. **Cassandra:** A Novel. 3 vols. Cr. 8vo, *cloth,* 31s. 6d.

Cox, Sir Geo. W., Bart., M.A. **The Little Cyclo-**pædia of Common Things. *Fourth Edition.* Illustrated. Demy 8vo, *cloth gilt,* 7s. 6d.

Craik, Georgiana M. **Twelve Old Friends.** With 8 Plates by Ernest Griset. Obl. 4to, *cloth gilt.* 5s.

Crawford, J. Coutts, F.G.S. **The Reform of Eng-**lish Spelling. Cr. 8vo, 6d.

Croker, T. Crofton. **Fairy Legends and Tra-**ditions of the South of Ireland. *New Edition.* [*Shortly.*

Cross—Davidson. **Stories of Great Men,** taken from Plutarch. By M. CROSS and A. J. DAVIDSON. Illustrated. Sm. 8vo, *cloth gilt, gilt edges,* 3s.

Cupples, Mrs. George. **Tappy's Chicks, and other** Links between Nature and Human Nature. With nineteen illustrations. Sm. 8vo, *cloth gilt, gilt edges,* 2s. 6d.

Dalton, Douglas. **False Steps:** a Novel. *Second Edition.* Cr. 8vo, 1s.

Daly, J. Bowles, LL.D. **Radical Pioneers of the** Nineteenth Century. Crown 8vo, *cloth,* 6s.

Darton. Books by J. M. DARTON:

Brave Boys who have become Illustrious Men of our Time. *Fourth Edition.* Plates. Cr. 8vo, *cloth gilt, gilt edges,* 3s. 6d. Thomas Carlyle — Robert Moffat — Professor Ruskin — George Cruikshank — John Stuckey Reynolds — Henry Deane, F.L.S. — William Chambers, and others.

Famous Girls who have become Illustrious Women of our Time. *Twentieth Edition.* Plates. Cr. 8vo, *cloth gilt, gilt edges,* 3s. 6d. Margaret Roper—"Little Miss Burney"—Laura Bridgman—Felicia Hemans — Harriet Beecher Stowe—Elizabeth Le Brun — Mme. de Staël — Frederika Bremer, and others.

De Portugall, Mme.   Synoptical Table of the Kindergarten.   Folio, mounted on canvas, and folding into a cloth case, 2s. 6d.

Dilke and Woodall.   Women Suffrage.   By Mrs. ASHTON DILKE and W. WOODALL, M.P.   Cr. 8vo, *cloth*, 1s.
[*Imperial Parliament Series.*

Douglas, W.   Measure for Measure : a Novel.   2 vols., cr. 8vo, 21s.

Dover.   Works by Rev. T. B. DOVER, M.A., Vicar of St. Agnes, Kennington.

Lent Manual : Some Quiet Lenten Thoughts. With a Preface by the BISHOP OF LINCOLN. Twelfth	Thousand. 12mo, 2s. 6d. Cheap edition, 1s. 6d. The Ministery of Mercy.

Eastward Ho!   A Monthly Magazine. Vols. I.—III., demy 8vo, *cloth*, each 3s. 6d.

Edwards, F. J.   Rules for the Mental Calculator. 12mo, *cloth*, 1s.

Edwards, R. O.   Rabbits for Exhibition, Pleasure, and Market.   By R. O. EDWARDS, assisted by several eminent breeders.   With eight plates.   Cr. 8vo, *limp cloth*, 2s. 6d.

Espin, Rev. T. H., M.A.   Elementary Star Atlas, with 12 large star-maps.   Imperial 8vo, *cloth*, 1s. 6d.

Evelac, Hamilton.   The Leaven of Malice : a Novel. Cr. 8vo, 6s.

Everitt, Graham.   English Caricaturists and Graphic Humourists of the Nineteenth Century.   Illustrated with a large number of woodcut reproductions of rare caricatures, book illustrations, etc.   4to, *cloth extra, gilt top,* 42s.

Ewing, R.   Handbook of Agriculture.   With Preface by Prof. John Scott.   12mo, *limp cloth*, 6d.

Fairy Library of All Nations.
SERIES I.   TALES OF THE PEOPLE (from oral tradition).
 Bowker's Goblin Tales of Lancashire, 2s. 6d.
 Caballero's Book of Spanish Tales, 2s. 6d.
 Cappel's Old Norse Sagas, 2s. 6d.
 Fryer's English Fairy Tales from the North Country, 2s. 6d.
 Geldart's Modern Greek Folklore, 2s. 6d.
 Gesta Romanorum, selected and adapted, 2s. 6d.
 Matthews' Hiawatha and other Legends of the Wigwams.   [*New Edition in preparation.*
 Rowsell's Spirit of the Giant Mountains, 2s. 6d.

[*See next page.*

Stephens' Old Norse Fairy Tales, 2s. 6d.

SERIES II. ORIGINAL FAIRY TALES.

Austin's Moonfolk, 2s. 6d.

Hauff's Popular Tales, 3s. 6d.

Parker's Among the Fairies, 2s. 6d.

Faithfull, Mrs. Century Cards: A New Method of teaching Chronology. In box, 10s.

Fawkes. Books by F. A. FAWKES, F.R.H.S.:

Horticultural Buildings, their Construction, Heating, Interior Fittings, etc., with Remarks on the Principles involved, and their application. With 123 cuts. *Second Edition*. Cr. 8vo, *cloth*, 3s. 6d.

Hot-water Heating. Cuts. 12mo, 1s.

Babies, and how to Rear them. Cr. 8vo, *limp cloth*, 6d.

Fillmore, J. C. A History of Pianoforte Music. Edited by Ridley Prentice. Roy. 16mo, *cloth*, 3s. 6d.

Fleay, F. G., M.A. The Logical English Grammar. Fcap. 8vo, *cloth*, 2s.

Forbes, Gordon S. Wild Life in Canara and Ganjam. With Coloured Plates. Cr. 8vo, *cloth*, 6s.

Fowle. Latin and Greek School Books by the Rev. EDMUND FOWLE, M.A.:

A New Latin Primer. Cr. 8vo. [*Shortly*.

A Short and Easy Latin Book. *New Edition*. Post 8vo, 1s. 6d.

A First Easy Latin Reading Book. *New Edition*. Post 8vo, 3s. 6d.

A Second Easy Latin Reading Book. *New Edition*. Post 8vo, 3s. 6d.

Selections from Latin Authors: Prose and Verse. Post 8vo, 2s. 6d. ; or in two Parts, 1s. 6d. each.

Short and Easy Greek Book. *New Edition*. Post 8vo, 2s. 6d.

First Easy Greek Reading Book. Containing Fables, Anecdotes of Great Men, Heathen Mythology, etc. *New Edition*. Post 8vo, 5s.

Second Easy Greek Reading Book. Containing Extracts from Xenophon, and the whole of the First Book of the Iliad. *New Edition*. Post 8vo, 5s.

First Greek Reader for Use at Eton. *New Edition*. Post 8vo, 1s. 6d.

The First Book of Homer's Iliad, in Graduated Lessons, with full notes and vocabularies. Post 8vo, 2s.

Friend. Works by Rev. HILDERIC FRIEND, F.L.S.

Flowers and Flower Lore. Illustrated. *Third Edition.* 8vo, *cloth gilt, gilt edges,* 7s. 6d.

The Ministry of Flowers. Illustrated. Crown 8vo, *cloth gilt, gilt top,* 2s. 6d.

Froebel, Friedrich. Selections from his Writings. Edited by H. K. Moore, B.A., and Mme. Michaelis.   [*Shortly.*

Fryer, Dr. A. C. Book of English Fairy Tales from the North Country. Plates. Cr. 8vo, *cloth gilt, gilt edges.* 2s. 6d.   [*Fairy Library.*

Fuller, Thomas, D.D.

Life of Thomas Fuller, D.D., the Church Historian. By the Rev. J. M. FULLER, M.D. *Second Edition.* 2 vols. Cr. 8vo, 12s.

Selections from the Holy and Profane States, with a Short Account of the Author and his Writings. Crown 8vo, 3s. 6d.

Gaussen. Works by Professor GAUSSEN. Cr. 8vo, each 1s. 6d.

The Iron Kingdom.
The King's Dream.

The Kingdom of Iron and Clay.

Geldart. Works by Rev. E. M. GELDART, M.A. :

Modern Greek Folklore. Cr. 8vo, *leatherette,* 2s. 6d.
Sunday for our Little Ones : Addresses to the Young. Cr. 8vo, *cloth, gilt edges,* 3s.

The Doctrine of the Atonement according to the Epistle of St. Paul. Cr. 8vo, *cloth,* 3s. 6d.
See also Zacher, p. 26.

Gems from the Poets. Illustrated with thirty coloured designs by A. F. Lydon. Imp. 8vo, *cloth extra, gilt edges,* 7s. 6d.

Geometry, Plane, The Elements of. Prepared by the Association for the Improvement of Geometrical Teaching. Part I. (corresponding to Euclid Bks. I.—II.) With numerous figures. Cr. 8vo, *cloth,* 3s. 6d.   [*Parts II. and III. at press.*

George II., History of the Reign of. By Oxon. Cr. 8vo, *cloth,* 3s.   [*Student's Manuals.*

George III., History of the Reign of. By Oxon (an Army Tutor). Based on Bright, Macaulay's Essays, Napier, Hughes, and Burke. To which are added 240 Examination questions. Cr. 8vo, *cloth,* 4s. 6d.   [*Student's Manuals.*

Gesta Romanorum. Selected and adapted. Plates. Cr. 8vo, *cloth gilt, gilt edges*, 2s. 6d.   [*Fairy Library.*

Gilbert. Books by WILLIAM GILBERT :

Modern Wonders of the World, or the New Sindbad. *Second Edition.* Illustrated by Arthur Hughes. Sm. 8vo, *cloth gilt, gilt edges*, 3s.

The History of a Huguenot Bible. *Second Edition.* Illustrated. Sm. 8vo, *cloth gilt, gilt edges*, 3s.

Goethe. Select Poems of Goethe, edited, with Introductions, Notes, and a Life of Goethe (in German), by Prof. E. A. Sonnenschein, M.A. (Oxon), and Prof. Alois Pogatscher. *Second Edition.* 12mo, *limp cloth*, 1s. 6d.   [*Annotated German Classics.*

Gorman, W. Gordon. Converts to Rome : a Classified List of nearly 4,000 Protestants who have recently become converted to the Roman Church. *Second Edition much enlarged* [1885]. Royal 16mo, *cloth extra, gilt top*, 2s. 6d.

Gronlund, L. The Co-operative Commonwealth. An Exposition of Modern Socialism. Cr. 8vo, *cloth*, 2s. ; *paper*, 1s.6d.

Gray, Peter. Lichens; Mosses, Scale Mosses, and Liverworts ; Seaweeds. With cuts, 12mo.   [*In the press.*

Greene, Rev. J. The Insect Hunter's Companion. *Third Edition.* Cuts. 12mo, *boards*, 1s.

Greenwood, James (the " Amateur Casual "). Reminiscences of a Raven. Illustrated. Fcap. 8vo, *cloth gilt*, 1s.

Grimm, Jacob. Teutonic Mythology, translated by J. Steven Stallybrass. 3 vols. Demy 8vo, *cloth*, 45s.

Guizot, F. The Devoted Life of Rachel, Lady Russell. Illustrated. Cr. 8vo, *cloth gilt, gilt edges*, 1s. 6d.

Gustafsson, Richard. Tea Time Tales for young Little Folks and young Old Folks. *Third Edition.* Illustrated. Cr. 8vo, *cloth gilt, gilt edges*, 3s. 6d.

Guyot, Yves. Hon. member of the Cobden Club. Principles of Social Economy. With numerous Diagrams. Demy 8vo, *cloth*, 9s.

Harley, Rev. Timothy. Moon Lore. Illustrated by facsimiles of old prints and scarce woodblocks. 8vo, *cloth extra, gilt top*, 7s. 6d.

Harris, Joel Chandler. Uncle Remus. Legends of the Plantations. *The Original Illust. Edition.* Cr. 8vo, *cloth*, 2s. 6 d

Harting, J. E.   Glimpses of Bird Life.   Illustrated with 20 coloured plates by P. Robert.   Royal folio, *cloth extra, gilt edges*, 42s.

Hauff, W.   Popular Tales.   Translated by Percy E. Pinkerton.   *New Edition*.   Cuts.   Cr. 8vo, *cloth gilt, gilt edges*, 3s. 6d.   [*Fairy Library*.

Hawthorne, Nathaniel.   Biographical Stories.   Portraits.   Cr. 8vo, *cloth gilt, gilt edges*, 1s. 6d.
Benj. West, Newton, Johnson, Cromwell, B. Franklin, Queen Christina.

Hawthorne, Dr. Robert.   The Student's Manual of Indian History.   Cr. 8vo, *cloth*, 3s. 6d.   [*Student's Manuals*.

Hehn, Prof. Victor.   The Wanderings of Plants and Animals.   Edited by J. Steven Stallybrass.   Demy 8vo, *cloth*, 16s.

Hein, Dr. G.   A German Copy-Book.   32 pages, each with a separate head-line, 4to, in wrapper, 6d.

Henderson, F. Leslie.   Three Plays for Drawing- Room Acting.   Cinderella, The Lady-Help, Story of the Stars.   Demy 8vo, 1s.

Hewett, H. G.   Heroes of Europe.   Illustrated.   Cr. 8vo, *cloth gilt, gilt edges*, 3s. 6d.

Hewetson.   Works by Dr. H. BENDELACK HEWETSON :

Life of Robert Hewetson.   Illustrated by phototypes.   Royal 4to, *boards*, 42s.
The Influence of Joy upon the Workman and his Work.
Illustrated by autotypes.   4to, *boards*, 3s. 6d.
The Human Eye in Perfection and in Error.   Cuts.   Demy 8vo, 1s.

Hichens, R. S.   The Coastguard's Secret : a Novel.   Cr. 8vo, *cloth*, 6s.

Higginson, T. Wentworth.   Common Sense about Women.   *Third Edition*.   Cr. 8vo, *cloth boards*, 1s.

Hillocks, Rev. J. Inches.   Hard Battles for Life and Usefulness.   With Introduction by Walter C. Smith, D.D., and photo. of the author.   *Second Edition*.   Demy 8vo, *cloth*, 3s. 6d.

Hinton, C. H., B.A.   Scientific Romances.   1s. each.
What is the Fourth Dimension ?   The Persian King.

Hobson. Works by Mrs. CAREY HOBSON:

The Farm in the Karoo. Illustrated. *Second Edition.* Crown 8vo, *cloth gilt, gilt edges,* 3s. 6d.

At Home in the Transvaal: or, Boers and Boers. 2 vols. Crown 8vo, *cloth,* 21s.

Howe, Cupples, Master Mariner. **The Deserted Ship.** A real story of the Atlantic. Illustrated by Townley Green. *Fourth Edition.* Sm. 8vo, *cloth gilt, gilt edges,* 2s. 6d.

Hughan, Samuel. Hereditary Peers and Hereditary Paupers: the two extremes of English Society. *Paper,* 1s.

Imperial Parliament Series. *See* page 27.

Impey, F. Three Acres and a Cow. With Preface by the RT. HON. J. CHAMBERLAIN, M.P., and Appendix by the DUKE OF ARGYLL. Cr. 8vo, *paper,* 6d.

Irving, Washington, The Beauties of. With 23 full-page plates by George Cruikshank. *Edition de luxe.* Imperial 32mo, *cloth extra, gilt top,* 2s. 6d.

Isocrates' Evagoras. Edited, with Introduction and Notes for the use of schools, by Henry Clarke, M.A. 12mo, *cloth,* 2s. 6d.

Jenkins, Edward, M.P. Jobson's Enemies. With 10 plates by F. Barnard. Cr. 8vo, *cloth,* 6s.

Jewitt, Llewellynn. Handbook of English Coins. With a Chapter on Greek Coins by Barclay V. Head (Brit. Mus.). Illustrated. Cr. 8vo, *cloth,* 1s.

Jones, C. A. The Saints of the Prayer-Book. 6 plates, royal 16mo, *cloth extra, gilt edges,* 2s. 6d.

Jung, Dr. K. Australia and her Colonies. Illustrated. Cr. 8vo, *cloth gilt, gilt edges,* 3s. 6d.

Karoly, Dr. Akin. The Dilemmas of Labour and Education. Cr. 8vo, *cloth,* 3s. 6d.

Keene, Katherine. Voiceless Teachers. Cuts. Cr. 8vo, *cloth gilt, gilt top,* 2s.

Kindergarten Manuals.

　Bülow's Child and Child Nature. 3s.

　Bülow's Handwork and Headwork. 3s.

　The Kindergarten: Essays on Principles and Practice [Froebel Society's Lectures]. 3s.

Kindergarten, The : Essays on Principles and Practice. Being a Selection of Lectures read before the London Froebel Society. Cr. 8vo, *cloth*, 3s.

Kirby. Works by W. F. KIRBY (Brit. Mus.) :

Handbook of Entomology. Illustrated with several hundred figures. 8vo, *cloth gilt*, 15s.
Evolution and Natural Theology. Crown 8vo, *cloth*, 4s. 6d.

Young Collector's Handbook of Entomology. Fully Illustrated. Cr. 8vo, *cloth*, 1s.

Kirton. Books by DR. J. KIRTON :

Happy Homes and How to Make Them. 104*th Thousand*. Cuts. 12mo, *cloth gilt, gilt edges*, 2s.

The Priceless Treasure : an Account and History of the Bible. *Fourth Edition*. Cuts. 12mo, *cloth gilt*, 2s.

Kroeker, Kate Freiligrath. Alice thro' the Look- ing-glass, and three other Plays for Children. Plates. Crown 8vo, *cloth gilt, gilt edges*, 2s. 6d.

Lamb, Charles and Mary. Mrs. Leicester's School. Illustrated. *New Edition*. Fcap. 8vo, *cloth gilt*, 1s.

Le Free, Richard. The History of a Walking Stick, in Ten Notches. Cr. 8vo, *cloth*, 6s.

Leith-Adams. Novels by MRS. LEITH-ADAMS. Cheap editions. Each vol. cr. 8vo, *cloth extra*, 3s. 6d.

Geoffrey Stirling.
Madelon Lemoine.

Cosmo Gordon. [*Shortly*.
Lady Deane. [*Shortly*.

Letters of the Martyrs. Selected and abridged. Portraits. Cr. 8vo, *cloth gilt, gilt edges*, 3s. 6d.
Letters of Cranmer, Ridley, Hooper, Taylor, Saunders, Philpot, Bradford, Whittell, Careless, Glover, Simson, and others.

Liefde, Jacob de. The Great Dutch Admirals. *Fifth Edition*. Illustrated by Townley Green. Cr. 8vo, *cloth gilt, gilt edges*, 3s. 6d.
Heemskerk, Hein, Marten Tromp, De With, De Ruyter, Evertsen, Cornelius Tromp.

Life at Home, at School, and at College. By an Old Etonian. Illustrated. Cr. 8vo, *cloth gilt, gilt edges*, 3s. 6d.

Little. Works by J. STANLEY LITTLE :

South Africa : A Sketch Book of men and manners. 2 vols. Demy 8vo, *cloth gilt, gilt top,* 21s.

What is Art? Cr. 8vo, *cloth,* 3s. 6d.

Little, Rev. H. W. A Short History of Russia Cr. 8vo, *paper,* 1s.

Locke, John. Essay on the Human Understanding. Book III. (On Words.) Edited by F. Ryland, M.A. Cr. 8vo, *cloth,* 4s. 6d.

Löfving. Works by CONCORDIA LÖFVING :

Physical Education, and its place in a rational system of education. Portrait. Cr. 8vo, *cloth.* 1s. 6d.

A Manual of Gymnastics. [*In preparation.*

Lorne, Marquis of, K.G., K.T. Imperial Federation. Cr. 8vo, *cloth,* 1s.   [*Imp. Parl. Series.*

Lubbock, Sir John, Bart., M.P. Representation. Cr. 8vo, *cloth,* 1s.   [*Imp. Parl. Series.*

Maccall, William. Christian Legends of the Middle Ages. Cr. 8vo, *cloth,* 3s. 6d.

McAlpine. Works by Professor D. McALPINE :

Life Histories of Plants. With an Introduction to the Comparative Study of Plants and Animals on a Physiological Basis. Illustrated. Roy. 16mo. [*In the press.*

Handbook of the Diseases of Plants. Illustrated. Demy 8vo. [*In preparation.*

McCarthy, Sergeant T. A. Quarterstaff : A Practical Manual. With Figures of the Positions. 12mo, *boards,* 1s.

Maitland, Agnes C. Madge Hilton; or, Left to Themselves. Illustrated. Cr. 8vo, *cloth gilt, gilt edges,* 2s. 6d.

Malins, J. Shakespearean Temperance Calendar. A Red-line Birthday Book. 16mo, *cloth gilt, gilt edges,* 2s. 6d.

Malleson, Mrs. Frank. Notes on the Early Training of Children. Cr. 8vo, *cloth,* 2s. 6d.

Marryat, Florence. Tom Tiddler's Ground. [*In May.*

Martineau des Chesney, Baroness. Marquise and Rosette, and the Easter Daisy. Illustrated. Sm. 8vo, *cloth gilt, gilt edges,* 3s.

Marvin. Works by CHARLES MARVIN :

Reconnoitring Central Asia. Adventures of English and Russian Explorers, Secret Agents and Special Correspondents in the Region between the Caspian and India from 1863 to 1884. With Illustrations and Map. *Second Edition.* Demy 8vo, *cloth gilt,* 7s. 6d.

Our Public Offices. *Third Edition.* Cuts. Cr. 8vo, *cloth,* 2s.

Matthews, C. **Hiawatha,** and other Legends from the Wigwams of the Red American Indians.
[New Edition in preparation.

Maynard, Rev. A. **Happy Wedded Life.** *New Edition.* Plates. 12mo, *cloth gilt,* 2s.

Meignan, Victor. **Over Siberian Snows.** Edited by William Conn. With 16 plates. Demy 8vo, *cloth gilt,* 16s.

Mentone, Guide to. By an Englishman. Folding Map. 12mo, *cloth,* 1s. 6d.

Miller, Rev. J. R., D.D. **The Perfect Home** Series. 5 vols., 12mo, *cloth gilt,* each 6d.
1. The Wedded Life.
2. The Husband's Part.
3. The Wife's Part.
4. The Parent's Part.
5. The Children's Part.

Milnes. Works by ALFRED MILNES, M.A. :

Problems and Exercises in Political Economy. Cr. 8vo, *cloth,* 4s. 6d.
[Student's Manuals.

Elementary Notions of Logic. *Second Edition. Enlarged.* 41 cuts. Crown 8vo, *cloth,* 2s. 6d.

Mongan, Roscoe, B.A. **Our Great Military Commanders.** Illustrated. Cr. 8vo, *cloth gilt, gilt edges,* 3s. 6d.
Marlborough—Clive—Wolfe—Wellington—The Crimean War—The Indian Mutiny—Wolseley—Gordon.

Montague, Colonel. **Dictionary of British Birds.** *New Edition.* Edited by E. Newman, F.L.S. Demy 8vo, *cloth gilt,* 7s. 6d.

Monteiro, H. **Tales of Old Lusitania,** from the Folk Lore of Portugal. Cr. 8vo, *cloth gilt, gilt top,* 3s. 6d.

Moore. Works by H. KEATLEY MOORE, B.Mus., B.A. :

The Child's Pianoforte Book. *Second Edition.* Illustrated by Kate Greenaway and others. Fcap. 4to, *cloth gilt,* 3s. 6d.

Music in the Kindergarten. 12mo, 4d.
See also FROEBEL.

Moore, Nina. Manual of Kindergarten Drawing. Plates. 4to, *cloth*, 3s. 6d.

Müller, Prof. Max. Deutsche Liebe (German Love). Fragments from the Papers of an Alien. Cr. 8vo, *vellum*, 5s. ; *cloth gilt*, 3s. 6d.

Mulley—Tabram. Songs and Games for Our Little Ones. By Jane Mulley. Music by M. E. Tabram. *Second Edition*. Cr. 8vo, 1s.

Naegeli—Schwendener. The Microscope: Theory and Practice. By Prof. C. Naegeli and Prof. S. Schwendener. With about 300 woodcuts. Demy 8vo, *cloth*, 21s. [*In the press.*

Naturalist's World, The. An Illustrated Monthly Magazine of Popular Science. 4to, vol. I. [1884], *cloth gilt*, 3s. [20 *pages, Monthly*, 2d.

Needlework for Ladies, for Pleasure and Profit. By " Dorinda." *Second Edition*. Crown 8vo, *boards*, 1s. 6d.

New Crusade, A. By Peter the Hermit. Illustrated. 8vo, *boards*, 2s.

Newman. Works by E. Newman, F.L.S. :

History of British Ferns. *Third Edition*. Cuts. Demy 8vo, *cloth*, 18s. A " People's Edition " of the same (abridged), containing numerous Figures, is also issued. *Fifth Edition*. 12mo, *cloth*, 2s. *See also* Montague's Dictionary of British Birds.

Nicholson, E. Student's Manual of German Literature. Cr. 8vo, *cloth*, 3s. 6d. [*Student's Manuals.*

Norton. Histories for Children, by Caroline Norton:

History of Greece. For children. 12mo. Illustrated. 1s.

History of Rome. For children. 12mo. Illustrated. 1s. History of France. For children. 12mo. Illustrated. 1s.

O'Reilly, Mrs. Robert. The Story of Ten Thousand Homes. *Second Edition*. Illustrated. Sm. 8vo, *cloth gilt*, *gilt edges*, 3s.

Orme, Temple (Teacher at University College School). The Rudiments of Chemistry. With several Woodcuts. Cr. 8vo, *cloth*, 2s. 6d.

Parker, Joseph, D.D. (of the City Temple). **Weaver** Stephen ; or, the Odds and Evens of English Religion. 8vo, *cloth*, 7s. 6d.

Parker, Hon. Mrs. Adamson. Among the Fairies. Illustrated by Lily Chitty. Cr. 8vo, *cloth gilt*, *gilt edges*, 2s. 6d. [*Fairy Library.*

Paul, Howard. Not too Funny, just Funny Enough ! Short Stories, American and Original. Cr. 8vo, *boards*, 1s.

Percy Reliques. The Reliques of Ancient Eng- lish Poetry, consisting of Old Heroic Ballads, Songs, and other Pieces. By THOS. PERCY, D.D., Bishop of Dromore. Edited, with an Introduction, Notes, and Glossary, by H. B. WHEATLEY, F.S.A. 3 vols., 8vo, *cloth extra*, 36s.

Perez, Bernard. The First Three Years of Child- hood. With a Preface by Prof. James Sully, M.A. Cr. 8vo, 4s. 6d.

Plautus' Captivi. Edited, with Introduction, Critical Apparatus and Notes, by Prof. E. A. Sonnenschein, M.A. (Oxon). Demy 8vo, *cloth*, 6s. School Edition of the same, with Notes. *Third Edition*. 3s. 6d.

Pooley—Carnie. The Common-Sense Method of Teaching French. By H. Pooley and K. Carnie. 12mo, *cloth*. Part I., 1s.; Part II., 1s. ; Memory Exercises, 1s. [*Other Parts in preparation.*

Prantl—Vines. Elementary Text Book of Botany. By Prof. W. Prantl and S. H. Vines, D.Sc., M.A., Fellow and Lecturer of Christ's College, Cambridge. *Fourth Edition* [1885]. 275 woodcuts, demy 8vo, *cloth*, 9s.

Prentice, Ridley. The Musician : A Guide for Pianoforte Students. In six Grades. Grades I.—IV. Roy. 16mo, *cloth*, *each* 2s. [*Other Grades in preparation.*

See also Fillmore's History of Pianoforte Music.

Ramsay, A., F.G.S. **Bibliography, Index and** Guide to Climate. Cuts. Demy 8vo, *cloth gilt*, 16s.

Rathbone and Pell. Local Government and Tax- ation. By W. RATHBONE, M.P., ALBERT PELL, M.P., and F. C. MONTAGUE, M.A. Cr. 8vo, *cloth*, 1s. [*Imp. Parl. Ser.*

Rawnsley, Rev. H. R. **Christ for To-Day**: A Series of International Sermons by Eminent Preachers of the English and American Episcopal Churches. Edited by Rev. H. R. RAWNSLEY, M.A., Vicar of Keswick. Imp. 16mo, *cloth, gilt top,* 6s.

Reid. Novels by Capt. MAYNE REID :

The Death Shot. Illustrated. Cr. 8vo, *cloth gilt, gilt edges,* 3s. 6d.
The Flag of Distress. Illus-

trated. Cr. 8vo, *cloth gilt, gilt edges,* 3s. 6d.
The Child Wife. Illustrated. Cr. 8vo. [*In preparation.*

Reynard the Fox. An old story new told. With Kaulbach's Illustrations. *Second Edition.* 4to, *cloth extra, gilt top,* 5s.

Rich, Elihu. **History of the War between** Germany and France, 1870-71. Fully Illustrated. Imp. 8vo, 21s.

Richard and Williams. **Disestablishment.** By HENRY RICHARD, M.P., and J. CARVELL WILLIAMS, M.P. *Second Edition.* Cr. 8vo, *cloth,* 1s. [*Imp. Parl. Ser.*

Richmond, the Rev. Legh. **Annals of the Poor.** With Memoir of the Author by J. S. Stallybrass. Plates. Cr. 8vo, *cloth gilt, gilt edges,* 1s. 6d.

" Robin." Children's Books by "ROBIN :"

The Little Flower Girl, and other Stories, in verse. Illustrated by Ernest Griset. Cr. 8vo, *cloth gilt, gilt edges,* 1s. 6d.
Skippo, and other Stories, in prose and verse. Illustrated by Ernest Griset. Cr. 8vo, *cloth gilt, gilt edges,* 1s. 6d.

Rogers. Works by Prof. J. E. THOROLD ROGERS, M.P. :

Six Centuries of Work and Wages : the History of English Labour. *Second Edition.* In 1 vol., 8vo, *cloth,* 15s.
Eight Chapters from the History of English Work and Wages, being a reprint of
certain chapters of "Six Centuries of Work and Wages." Crown 8vo, *cloth,* 3s. 6d.
Ensilage, and its Prospects in English Agriculture. *Second Edition.* Cuts. Cr. 8vo, *limp cloth,* 1s.

Rooper. Books by W. and H. ROOPER :

An Illustrated Manual of Object Lessons, containing hints for Lessons in Thinking and Speaking, with 20 "blackboard " illustrations. Cr. 8vo, *cloth,* 3s. 6d.
A Manual of Collective Lessons in Plain Needlework and Knitting. With numerous Plates and Diagrams. Cr. 8vo, *cloth,* 3s. 6d.

Ross, Ellen (Author of " The Candle Lighted by the Lord "). Dora's Boy. Fifth Thousand. Illustrated. Small 8vo, *cloth gilt, gilt eages,* 3s

Rouse, Lydia L. Sandy's Faith. A tale of Scottish Life. Illustrated. *Second Edition.* Fcap. 8vo, *cloth gilt,* 1s.

Rowe. Tales by RICHARD ROWE :

Roughing it in Van Diemen's Land, and Harry Delane. Sm. 8vo, *cloth gilt, gilt edges,* 3s.

A Haven of Rest, and Dr. Pertwee's Poor Patients. Sm. 8vo, *cloth gilt, gilt edges,* 3s.

Rowsell. Books by MARY C. ROWSELL :

Sweet Bells Jangled : A Novel. 3 vols. [*At press.* Tales of Filial Devotion. Illustrated. Cr. 8vo, *cloth gilt, gilt edges,* 2s. 6d.

The Spirit of the Giant Mountains. Illustrated. Cr. 8vo, *cloth, gilt edges,* 2s. 6d. [*Fairy Library.*

Rye, John, M.A. **Kirby in the Dale.** A Novel. 3 vols. Cr. 8vo, *cloth,* 31s. 6d.

Schaible—Althaus. Seeing and Thinking : Elementary Lessons and Exercises, introductory to Grammar, Composition, and Logical Analysis. By C. H. SCHAIBLE, M.D., F.C.P., and T. H. ALTHAUS, M.A., Oxon. *Second Edition.* Cr. 8vo, *cloth,* 3s. 6d.

Schiller's Cabal and Love. Translated by T. S. Wilkinson. 12mo, *leatherette,* 2s. 6d.

Scottish Naturalist, The. Demy 8vo. Quarterly, 1s. 2d.

Scott, Redna. Edith : A novel. 3 vols. Cr. 8vo, 31s. 6d.

Seboth, J. Alpine Plants. Painted from Nature, with descriptive text by A. W. Bennett, M.A., B.Sc. 4 vols. each with 100 coloured plates. Super roy. 16mo, *half persian, gilt tops, each* 25s.
The whole series (four vols.) in an elegant carved cabinet, £6 6s. *nett.*

Shakespeare. The Works of WILLIAM SHAKESPEARE. The Text revised by Rev. ALEXANDER DYCE. In 10 volumes, 8vo, with Life, Portraits, Facsimile of Will, etc. *Fifth Edition. Beautifully printed on antique-laid paper, and handsomely bound in cloth extra, gilt top,* each vol, 9s. [*Vols. I.—VI. ready.*

Shakspere. Othello. Edited for School Use, with notes, by Roscoe Mongan, B.A. Royal 16mo, *cloth,* 2s.

Shakspere, The Life and Times of. Portraits. Cr. 8vo, *cloth gilt, gilt edges,* 1s. 6d.

Sherwood's, Mrs. Juvenile Library. In three
series. Cuts. 12mo, *cloth gilt, each*, 1*s.*

Shields, Rev. R. J. Knights of the Red Cross:
Seven Allegorical Stories. Plates. 12mo, *cloth gilt*, 1*s.*

Shilling Gift Books. Illustrated. Fcap. 8vo, *cloth, gilt.*

Mrs. Leicester's School. By **Charles** and **Mary**
Lamb.

Sandy's Faith. A Tale of Scottish Life. By **Lydia**
L. Rouse.

The Knights of the Red Cross : Seven Allegorical
Stories. By the Rev. R. J. Shields.

Crimson Pages. A Tale of the Reformation. By W.
Tillotson.

Reminiscences of a Raven. By James Greenwood
(the "Amateur Casual").

Shirreff. Kindergarten Books by EMILY A. SHIRREFF :

The Kindergarten: Principles of Froebel's System, and their bearing on the Education of Women. *Third Edition.* Cr. 8vo, *cloth*, 1*s.* 4*d.*	The Kindergarten and the School. 12mo, 3*d.* Wasted Forces. 12mo, 3*d.*

Sime. Novels by WILLIAM SIME.

The Red Route. 3 vols. 31*s.* 6*d.* | Cradle and Spade. 3 v. 31*s.* 6*d.*

Sixpenny Gift Books. Illustrated. Demy 32mo, *cloth gilt.*

1. Little Henry and his Bearer.	7. Little Goody Two-Shoes.
2. Cheerful Cherry; or, Make the Best of it.	8. Little Dickie : a Simple Story.
	9. Three Foolish Little Gnomes.
3. The Basket of Flowers.	10. Cat and Dog Stories.
4. The Babes in a Basket.	11. Story of Patient Griseldis.
5. The Prince in Disguise.	12. Language of Flowers.
The Wanderer.	

Solly, the Rev. Henry. Rehousing of the Industrial Classes, or Village Communities *v.* Town Rookeries. 16mo, *limp cloth*, 6d.

Sonnenschein's Three Shillings and Sixpenny Novels.

Abdy-Williams, E. M. Two Ifs.

Churchill, Mrs. Spencer. Daisy Darling.

Leith-Adams, Mrs. Geoffrey Stirling.

„   „   Madelon Lemoine.

Mayne Reid, Capt. The Death Shot.

„   „   The Flag of Distress.

Tytler, C. C. Fraser. Jasmine Leigh.

„   „   Margaret.

Williams, Sarah ("Sadie"). The Prima Donna.

Sonnenschein—Nesbitt. Arithmetical works by A. Sonnenschein and H. A. Nesbitt, M.A. :

The Science and Art of Arithmetic. Part I., 2s. 6d. ; Parts II.—III., 3s. 6d. ; Parts I.—III. in one vol., 5s. 6d. Exercises (only), Part I., 1s. ; Parts II.—III., 1s. 3d. Answers (complete), 1s. 6d.

ABC of Arithmetic. Teacher's Book. Part I., 1s. ; Part II., 1s. Pupil's Book (Exerc. only), Part I., 4d. ; Part II., 4d. Ciphering Book. 40pp. chequered on right-hand page, and ruled on left-hand page for teacher's remarks. 3s. *per doz.*

Sonnenschein, A. Foreign Educational Codes relating to Elementary Education, prescribed by Austrian, Belgian, German, Italian, and Swiss Governments, with Introduction and Notes. Cr. 8vo, *cloth*, 3s. 6d.

Sonnenschein's Number Pictures. Fourteen folio coloured sheets for teaching the rudiments of number. *Fifth Edition.* On *one roller*, 7s. 6d.; on *boards varnished*, 16s. Descriptive pamphlet, 6d.

Sonnenschein's Patent Arithmometer. Box *a*, 5s. 6d.; box *b*, 4s. 6d. ; box *c*, 20s. Complete set, £1 10s.

Sonnenschein's Special Merit Readers. Each well and fully illustrated, and strongly bound in *cloth*. Parts I.—II. *at press.* Part III. (Standard III.), 1s. Part IV. (Standard IV.), 1s. 4d.

Sonnenschein's Linear Blackboard (Outline)
Maps (rolling up).

England and Wales. 4 ft. 9 in. by 4 ft. 16s.	Two Hemispheres. 5 ft. 6 in. by 4 ft. 21s. [Shortly.
Europe. 5 ft. 6 in. by 4 ft. 21s.	Others in preparation.

Stafford, Eric. Only a Drop of Water and other tales. *Third Edition.* Illustrated. Cr. 8vo, *cloth gilt, gilt edges,* 1s. 6d.

Stephens, George. Old Norse Fairy Tales. Cuts. Cr. 8vo, *cloth gilt, gilt edges,* 2s. 6d. [*Fairy Library.*

Stories of my Pets. Illustrated. Cr. 8vo, *cloth, gilt edges,* 1s. 6d.

Strong and Meyer. A History of the German Language. By H. A. STRONG, Professor of Latin in the Liverpool University College; and KUNO MEYER, Lecturer on Teutonic Languages, Liverpool University College. 8vo, *cloth,* 6s.

Stubbs. Works by the REV. CHARLES W. STUBBS, M.A. :

Christ and Democracy. Cr. 8vo, *cloth gilt,* 3s. 6d.	The Conscience and other Poems. Printed on hand-made paper. 12vo, *vellum,* 2s. 6d.
The Land and the Labourers. *Second edition.* Cr. 8vo, 1s.	Anthology of Christian Morals. [*In preparation.*

Student's Manuals.

Hawthorne's Student's Manual of Indian History, 3s. 6d.

Milnes' Problems and Exercises in Political Economy, 4s. 6d.

Nicholson's Student's Manual of German Literature, 3s. 6d.

"Oxon's" Student's Manual of the Reign of George III., 4s. 6d.

Table Books.

The Graphic Table Book. 1d.; *cloth,* 2d.

The Eclipse Table Book. 130th thousand, ½d.

Taylor, Jeremy. Selections from the Works of. With a Short Account of the Author and his Writings. 3s. 6d.

Theal, George McCall. Kaffir Folk Lore; with an Introduction on the Mythology, Manners, and Customs of the Kaffirs. *Second Edition.* Cr. 8vo, *cloth gilt, gilt top*, 4s. 6d.

Thorn. Coloured books with text by ISMAY THORN :

Harlequin Eggs. A 4to colour-book for children, with 24 pages of pictures by Lily Chitty. *Illustrated boards*, 2s. 6d.

In and Out. A 4to colour-book for children, with 24 pages of pictures by Lily Chitty. *Illustrated boards*, 2s. 6d.

Tillotson, W. Crimson Pages : a Story of the Reformation. Plates. 12mo, *cloth gilt*, 1s.

Time. A Monthly Magazine of Current Topics, Literature and Art. Medium 8vo.
  Vols. I.—IX., edited by EDMUND YATES. £3.
  Vols. X.—XI. (1884), edited by B. MONTGOMERIE RANKING. Each 6s.
  New Series, edited by E. M. ABDY-WILLIAMS, commencing with January, 1885. Vols. 1—2, each, 7s. 6d.
                [Monthly, 1s.

Tiny Mite, the Adventures of a Little Girl in Dreamland. With a large number of Illustrations. 4to, *cloth*, 5s.

Turner, F. C., B.A. A Short History of Art. Illustrated. Demy 8vo, *cloth gilt, gilt top*, 12s. 6d.

Tytler. Novels by C. C. FRASER TYTLER :

Jasmine Leigh. *Second Edition*, 3s. 6d.

Margaret. *Sec. Edition*. 3s. 6d.
Jonathan. *Sec. Edition*. [*Shortly*.

Tytler, M. Fraser. Tales of many Lands. Illustrated. Sm. 8vo, *cloth gilt, gilt edges*, 3s.

Valvedre, A. de. Sorrowful yet Lucky. A Novel. 3 vols. Cr. 8vo, *cloth*, 31s. 6d.

Vernalecken, Th. In the Land of Marvels. Folk tales of Austria and Bohemia. Edited by the Rev. Prof. E. Johnson, M.A. Cr. 8vo, *cloth, gilt top*, 5s.

Vicary, J. Fulford, J.P. Readings from the Dane : Short Stories translated from contemporary Danish writers. Cr. 8vo, *paper*, 1s.

Villari, Lina. Life in a Cave. Frontispiece. Cr. 8vo, *cloth gilt, gilt edges*, 1s. 6d.

Vines, S. H., D.Sc., M.A. **A School Botany.**
[*In preparation.*]
See also Prantl—Vines.

Wägner. Works by Dr. W. WÄGNER :

Asgard and the Gods. A Manual of Norse Mythology. *Third Edition.* Illustrated. Demy 8vo, 7*s*. 6*d*.

Epics and Romances of the Middle Ages. *Second Edition.* Illustrated. Demy 8vo, 7*s*. 6*d*.

Wallace, Cornelia. Flowers, a fantasy. With miniature illustrations. Demy 32mo, *cloth gilt, gilt edges*, 6*d*.

Wallis. Novels by A. S. C. WALLIS :

In Troubled Times. A Novel. Translated from the Dutch by E. J. Irving. *Third Edition* (re-translated). Cr. 8vo, 6*s*.

Royal Favour. A novel. Translated from the Dutch by E. J. Irving. *Second edition.* Cr. 8vo, *cloth*, 6*s*.

"Wanderer" (Author of "Fair Diana," "Across Country," etc.). **Glamour: a Novel.** 3 vols. Cr. 8vo, 31*s*. 6*d*.

Weir. Works by ARCHIBALD WEIR, B.A. :

The Historical Basis of Europe. 8vo. [*Shortly.*]

The Critical Philosophy of Kant. Cr. 8vo, 2*s*. 6*d*.

Welby, S. E. The Traveller's Practical Guide. In four languages. A waistcoat pocket volume. *Cloth*, 1*s*.; *roan*, 1*s*.6*d*.

What the Boy thought. A social satire. Sixth thousand. Roy. 16mo, *parchment wrappers*, 6*d*.

White. Books by F. A. WHITE, B.A. :

An Unconventional English Grammar. *Second Edition.* 12mo, *cloth*, 4*s*.

The Boys of Raby. A holiday book for boys. Illustrated by J. Dinsdale. Cr. 8vo, *cloth gilt, gilt edges*, 2*s*. 6*d*.

Wiebe, Prof. E. The Paradise of Childhood: A complete manual of Kindergarten instruction. *Third Edition.* 75 plates. 4to, *cloth*, 10*s*. 6*d*.

Williams, Sarah ("Sadie"). The Prima Donna. A Novel. 1 vol. edition. Cr. 8vo, *cloth*, 3*s*. 6*d*.

Wilson, Rev. John M. Nature, Man, and God. Contributions to the Scientific Teaching of To-day. Cr. 8vo, *cloth*, 5s.

Wood, Rev. H. A Season among the Wild Flowers. *Second Edition.* Cuts. Cr. 8vo, *cloth gilt, gilt edges*, 3s. 6d.

Wright, Dr. Alfred. Adventures in Servia: Experiences of a Medical free-lance among the Bashi-Bazoucs, etc. Edited and illustrated by E. Farquhar-Bernard, M.R.C.S. (*late Surgeon of the Servian Army*). Demy 8vo, *cloth gilt*, 10s. 6d.

Wurtz, Dr. A.   The Elements of Modern Chemistry. Cuts. Cr. 8vo, *cloth*, 10s. 6d.

Xenophon. The Hiero. Edited, with Introduction and Notes for the use of schools, by R. Shindler, M.A. Interleaved. 12mo, *cloth*, 2s. 6d.

Yonge. Biographical Books by Professor C. D. Yonge:

The Seven Heroines of Chris-tendom. *Third Edition* Illustrated. Cr. 8vo, *cloth gilt, gilt edges*, 3s. 6d.	**Our Great Naval Command-**ers. Illustrated. Cr. 8vo, *cloth gilt, gilt edges*, 3s. 6d. Drake — Blake — Cook — Rodney — Nelson — Parry.

Youthful Nobility. Plates. Cr. 8vo, *cloth gilt, gilt edges*. 1s. 6d.

Zacher, Dr. B. (Assessor to the Prussian Government). The Red International : An Account of Modern Socialism in Germany, France, Great Britain, Ireland, Switzerland, Belgium, Holland, Denmark, Scandinavia, Spain, Portugal, Italy, Austria, Russia, and North America. Translated by the Rev. E. M. Geldart, M.A. Cr. 8vo, *paper*, 1s.

Zimmern, Helen. Tales from the Edda. Illustrated by Kate Greenaway and others. Cr. 8vo, *cloth gilt, gilt edges*, 1s. 6d.

The Imperial Parliament Series.

Written entirely by MEMBERS OF PARLIAMENT. Edited by SYDNEY BUXTON, M.P.

In Uniform Crown 8vo Volumes, red cloth, neat, each about 150 pp. 1s.

1. **Marq. of Lorne.** Imperial Federation.
2. **Sir J. Lubbock.** Representation.
3. **W. Rathbone, Alb. Pell,** and **F. C. Montague.** Local Government and Taxation.
4. **Rt. Hon. W. E. Baxter.** England and Russia in Asia.
5. **Mrs. Ashton Dilke** and **W. Woodall.** Women Franchise.
6. **W. S. Caine, W. Hoyle,** and **Rev. Dawson Burns.** Local Option.
7. **Henry Broadhurst** and **R. T. Reid.** Leasehold Enfranchisement.
8. **Henry Richard** and **Carvell Williams.** Disestablishment.
9. **J. Bryce.** The House of Lords.
10. **J. F. B. Firth.** London Government and City Guilds.

The last two not yet ready. Others to follow.

Historical, Political and Social Science.

Daly's Radical pioneers, 6s.
Gronlund's Co-op. commonwealth, 2s.
Guyot's Social economy, 9s.
Higginson's Common sense about women, 1s.
Hughan's Hereditary peers and hereditary paupers, 1s.
Imperial Parliament Series, p. 27.
Karoly's Dilemmas of labour, etc., 3s. 6d.

Milnes' Political economy, 4s. 6d.
Rogers' Six centuries of work and wages, 15s.
,, Eight chapters from the history of English work and wages, 3s. 6d.
Solly's Rehousing the poor, 6d.
Stubbs' Christ and democracy, 3s. 6d.
,, Land and the labourers, 3s. 6d.
Zacher's Red international, 1s.

Agriculture, etc.

Chitty's Beekeeping.
Edwards' Rabbits, 2s. 6d.
Ewing's Agriculture, 6d.

Fawkes' Horticultural buildings, 3s. 6d.
,, Hot water heating, 1s.
Rogers' Ensilage, 1s.

Natural History and Science.

"Alert," Cruise of the, by Coppinger, 6s.
Allen's (Grant) The Evolution of Flowers.
Alpine Plants, 400 coloured plates, 4 vols., £5. In Cabinet, £6 6s.
Bennett's Flora of Alps, 5s.
Bevan's Royal Relief Atlas, 21s.
Claus-Sedgwick's Text-book of zoology, Vol. I., 21s. ; Vol. II., 16s.
Cox's Little cyclopædia of common things, 7s. 6d.
Espin's Star atlas, 1s. 6d.
Friend's Flowers and flower-lore, 7s. 6d.
Harting's Glimpses of bird life, 42s.
Hehn's Wanderings of plants, 16s.

Hewetson's The human eye, 1s.
Kirby's Handbook of entomology, 15s.
,, Evolution and nat. theology, 4s. 6d.
McAlpine's Diseases of plants.
,, Life histories of plants.
Montague's Dictionary of British birds, 7s. 6d.
Naegeli-Schwendener's The microscope, 21s.
Newman's History of British ferns, 18s.
Prantl-Vines' Text-book of botany, 9s.
Ramsay's Bibliography of climate, 16s.
Wurtz's Elements of modern chemistry, 10s. 6d.

POPULAR SCIENCE.

Armstrong's Birds and their ways, 1s. 6d.
Bagnall's Mosses, 1s.
Bottone's The dynamo, 2s. 6d.
Butler's Entomology of a pond.
Greene's Insect hunter's companion, 1s.
Clement's Geological systems, 1s.

Kirby's Young collector, 1s.
Newman's Ferns, People's edition, 2s.
Pilter's Human physiology, 1s.
Wood's Season among wild flowers, 2s. 6d.
Young Collector's Penny handbooks, 8 vols., 1d. each.

NATURAL HISTORY MAGAZINES.

The Naturalist's World. Monthly, 2d. | **The Scottish Naturalist,** quart., 1s.

Books of Travel, etc.

"Alert," Cruise of H.M.S. *Alert*, 6s.
Barras' India, 4 vols., each 3s. 6d.
Forbes' Canara and Ganjam, 6s.
Hobson's The farm in the Karoo, 3s. 6d.
Little's South African sketch book, 2 vols., 21s.

Marryat's Tom Tiddler's ground.
Marvin's Reconnoitring Central Asia, 7s. 6d.
Meignan's Over Siberian Snows, 16s.
Wright's Adventures in Servia, 10s. 6d.

Novels and Minor Fiction.

Abdy-Williams' Two Ifs, 3s. 6d.
   „   For his friend, 3 v., 31s. 6d.
   „   Forewarned, 1s.
Auerbach's Two stories, 2s. 6d.
Babcock's Cypress Beach, 2 vols., 12s.
Bickerdyke's With the best intentions, 2s. 6d.
Boger's Elfrica, 3 vols., 31s. 6d.
Bradshaw's Roger North, 3 vols., 31s. 6d.
Burke's Couleur de rose, 2 vols., 21s.
Churchill's Daisy Darling. 3s. 6d.
Corbett's Cassandra, 3 vols., 31s. 6d.
Dalton's False steps, 1s.
Douglas' Measure for measure, 2 vols., 21s.
Evelac's Leaven of malice, 6s.
Hichens' Coastguard's secret, 6s.
LeFree's Walking stick, 6s.
Hobson's At home in the Transvaal, 2 vols., 21s.

Leith-Adams' (Mrs.) Geoffrey Stirling, 3s. 6d.
Leith-Adams' Madelon Lemoine, 3s. 6d.
Mayne Reid, The Death Shot, 3s. 6d.
   „   „   The Flag of Distress, 3s.6d.
Müller's German love, 3s. 6d. and 5s.
Parker, Dr. J. Weaver Stephen, 7s. 6d.
Paul's Not too funny ! 1s.
Rowsell's Sweet bells jangled, 3 vols.
Rye's Kirby in the dale, 3 vols., 31s. 6d.
Scott's (Redna) Edith, 3 vols., 31s. 6d.
Sime's The red route, 3 vols., 31s. 6d.
Tytler's Jasmine Leigh, 3s. 6d.
   „   Margaret, 3s. 6d.
Valvedre's Sorrowful yet lucky, 31s. 6d.
Vicary's Reading from the Dane, 1s.
Wallis' In troubled times, 6s.
   „   Royal favour, 6s.
"Wanderer's" Glamour, 3 vols., 31s. 6d.
Williams' (S.) Prima donna, 3s. 6d.

Antiquities, Folk-lore, etc.

Axon's Gipsy folk tales.
Bowker's Goblin tales of Lancs., 2s. 6d.
Caballero's Book of Spanish tales, 2s. 6d.
Cappel's Old Norse sagas, 2s. 6d.
Croker's Irish fairy legends.
Friend's Flowers and flower-lore, 7s. 6d.
Fryer's English fairy tales, 2s. 6d.
Geldart's Modern Greek folk-lore, 2s. 6d.
Gesta Romanorum, 2s. 6d.
Grimm's Teutonic mythology, 3 vols.,45s.
Harley's Moon lore, 7s. 6d.
Harris' Uncle Remus, 2s. 6d.
Hehn's Wanderings of plants, 16s.

Maccall's Christian legends, 3s. 6d.
Matthews' Legends of the wigwams.
Monteiro's Portuguese folk-lore, 3s. 6d.
Percy Reliques, 3 vols., 31s. 6d.
Rowsell's The Spirit of the Giant mountains, 2s. 6d.
Stephens' Old Norse fairy tales, 2s. 6d.
Theal's Kaffir folk-lore, 4s. 6d.
Vernalecken's In the land of marvels, 5s.
Wägner's Asgard and the gods, 7s. 6d.
   „   Epics and romances, 7s. 6d.
Zimmern's Tales from the Edda, 1s. 6d.

Theological and Devotional Books.

Adams' My man and I, 7s. 6d.
Church Rambles and Scrambles, 2s.
Cockburn, Laws of nature, 3s. 6d.
Contemporary Pulpit. Vols. I.—IV., each, 6s. *Monthly*, 6d.
Dover's Lent manual, 2s. 6d. and 1s. 6d.
   „   Ministery of mercy, 6s.
Fuller's Holy and profane states, 3s. 6d.
   „   Life of Fuller, 2 vol., 12s.
Geldart's Sunday for our little ones, 3s.
   „   Doctrine of atonement, 3s. 6d.

Gorman's Converts to Rome, 2s. 6d.
Kirby's Evolution and nat. theol., 4s. (d.
Maccall's Christian legends, 3s. 6d.
Miller's The perfect home. 5 vols., ea. 6d.
Rawnsley's Christ for to-day, 6s.
Richard and Williams' Disestablishment, 1s.
Stubbs' Christ and democracy, 3s. 6d.
   „   Anthology of Christian morals.
Taylor, Jeremy, Selections from, 3s. 6d.
Wilson's The Supreme Power.

Temperance and Cottage Books.

Arthur's Ten nights, 2s.
Eclipse Elocutionist, 1s.
Kirton's Happy homes, 2s.
,,　Priceless treasure, 2s.
Malin's Shakespeare temp. cal., 2s. 6d.
Maynard's Happy wedded life, 2s.

Miller's The perfect home. 5 vols., each 6d.
Prize Pictorial Readings, 2s.
Rainbow Readings, 1s.
Sixpenny Series.
Wheeler's Drops of water, 1s.

Books on and of Music.

Andersen's Fairy tales set to music, 1s.6d
Bourne's Great composers, 3s. 6d.
Fillmore's Hist. of pianoforte music, 3s. 6d.
Moore's Child's pianoforte book, 3s. 6d.

Moore's Music in the K. G., 4d.
Mulley's Songs and games, 1s.
Pag1's Number notation, 1s. 6d.
Prentice's Musician, Grades I.—IV., 2s. each.

Books on the Fine Arts, etc.

Alpine Plants, 4 vols., each 25s.
Everitt's English caricaturists, 42s.
Harting's Glimpses of bird life, 42s.
Hewetson's Life of Hewetson, 42s.
Hewetson's Influence of joy, 3s. 6d.

Irving (Wash.), Beauties of. 23 plates by G. Cruikshank, 2s. 6d.
Little's What is art? 3s. 6d.
Turner's Short history of art, 12s. 6d.

Kindergarten Books.

Buckland's Happiness of childhood, 6d.
,,　Use of stories, 3d.
Bülow's Child nature, 3s.
,,　Hand-work and head-work, 3s.
DePortugall's Synoptical table, 2s. 6d.
Froebel, Selections from.
Heerwart's Mutterlieder, 3d.
Kindergarten, The. Essays, etc., 3s.

Moore's (H.K.) Child's pianof. book, 3s.6d
,,　,,　Music in the K.G., 4d.
,,　(N.) Kindergarten drawing, 3s.6d.
Mulley's Songs and games, 1s.
Shirreff's The Kindergarten, 1s. 4d.
,,　Wasted forces, 3d.　[3d.
,,　The Kinderg. and the School,
Wiebe's Paradise of childhood, 10s. 6d.

Books on Education.

Buxton's Overpressure, 2s. and 1s.
Cooper's Training of the memory, 3d.
Crawford's Reform of spelling, 6d.
Fawkes' Babies ; how to rear them, 6d.
Hoggan's Physical education of girls, 4d.
Karoly's Dilemmas of labour and education, 3s. 6d.
Kindergarten Books. *See* special heading above.
Locke "On words," ed. Ryland, 4s. 6d.
Löfving's Physical education, 1s. 6d.

Löfving's Manual of gymnastics.
McCarthy's Government code, 6d.
Malleson's Early training of children, 2s. 6d.
Moore's Selections from Froebel.
Nicholson's Student's manual of German literature, 3s. 6d.
Perez's First three years of childhood, 4s. 6d.
Sonnenschein's Foreign educational codes, 3s. 6d.

School and College Books, etc.

Bevan's Royal relief atlas, 21s.
,,　Home geography.
Brant's Systematic cutting out, 2s.

Claus—Sedgwick. Elem. Text-book of Zoology, 21s. and 16s.
Edwards' Mental calculator, 1s.

Faithfull's Century cards.
Fleay's Logical English grammar, 2s.
Fowle's Short and easy Latin book, 1s. 6d.
,,   First easy Latin reader, 3s. 6d.
,,   Second easy Latin reader, 3s. 6d.
,,   Short and easy Greek book, 2s. 6d.
,,   First easy Greek reader, 5s.
,,   Second easy Greek reader, 5s.
,,   First Greek reader for Eton, 1s. 6d.
,,   First book of Homer's Iliad, 2s.
,,   Selections fr. Lat. authors, 2s. 6d. and 1s. 6d.
Geometry, Plane, Elements of, 3s. 6d.
George II., 3s.   George III., 4s. 6d.
Goethe, Select poems of, 1s. 6d.
Hawthorne's Manual of Indian history.
Hein's German copy book, 6d.   [3s. 6d.
Isocrates, Evagoras, ed. Clarke, 2s. 6d.
Limerick, Bishop of. Geomet. models.
Milnes' Political economy, 4s. 6d.
,,   Elementary notions of logic, 2s. 6d.
Moore's Child's pianoforte book, 3s. 6d.
Norton's Histories, 3 vols., 1s. each.
Orme's Chemistry, 2s. 6d.
Pilter's Human physiology, 1s.

Plautus, The captivi, 6s., 3s. 6d.
Pooley-Carnie's Com. sense French, 1s.
Prantl-Vines' Text-book of botany, 9s.
Prentice's Musician. Grds. I.—IV. ea. 2s.
Rooper's Manual of object lessons, 3s. 6d.
,,   Needlework and knitting, 3s. 6d.
Schaible-Althaus' Seeing and thinking, 3s. 6d.
Shakspere's Othello, for school use, 2s.
Sonnenschein's—
   Number pictures, 7s. 6d. and 16s.
   Blackboard maps, 16s. and 21s.
   Patent arithmometer, 5s. 6d., 4s. 6d., and 30s.
   Special merit readers, 1s. and 1s. 4d.
   Science and art of arithmetic, 2s. 6d., etc.
   A B C of Arithmetic, 1s., etc.
   Ciphering book, 3s. per dozen.
Strong's History of German Lang., 6s.
Student's Manuals.
Table Books, ½d. and 1d.
Vines' School botany.
White's Unconventional Engl. gram., 4s.
Xenophon's Hiero, ed. Shindler, 2s. 6d.

The Cambridge Examiner, Monthly 6d.

Miscellaneous and Reference Books.

Best Books, The, a classified Bibliography.
Bevan's Guide to Westminster Abbey.
,,   ,,   Lichfield Cathedral.
Cox's Little cyclopædia of common things,
Jewitt's English Coins, 1s.   [7s. 6d.

McCarthy's Quarterstaff, 1s.
Marvin's Our public offices, 2s.
Mentone, Guide to, 1s. 6d.
Welby's Traveller's pract. guide, 1s.
What the boy thought, 6d.

GIFT AND PRIZE BOOKS.

Book at £6 6s. (nett.)

Seboth and Bennett's Alpine plants. 4 series, in cabinet.

Books at £1 1s.

Bevan's Royal Relief Atlas.

Rich's The war between Germany and France 1870-71.

Gift Books at 7s. 6d.

Adams' Book of earnest lives.
,,   Battle stories.
,,   Girlhood of remarkable women.
Cox's Little cyclopædia of common things.
Friend's Flowers and flower-lore.

Gems from the Poets.
Wägner's Asgard and the gods.
,,   Epics and romances of the Middle Ages.

Gift Book at 6s.

Alert. Cruise of the *Alert*, by Coppinger.

Gift Books at 5s.

Müller's (Max), German Love, *vellum*.
Reynard the Fox.

Vernalecken's In the land of marvels.

Gift Books at 3s. 6d.

Adams' Celebrated women travellers.
Alberg's Gustavus Vasa.
Arabian Nights, the new.
Birthday Book, Floral.
Bourne's The great composers.
,,   African heroes. 2 series.
Chapman's Notable women of the Re-
　　formation.
　　,,   Notable women of the
　　Puritan times.
Clarke's Book of Golden Friendships.
Cooke's Three great lives.
Darton's Brave boys.
,,   Famous girls.
Fuller's Holy and profane states.
Gustafsson, Tea-time tales.

Hauff's Popular Tales.
Hewett's Heroes of Europe.
Hillocks' Hard batiles.
Hobson's The farm in the Karoo.
Jung's Australia and her colonies.
Letters of the Martyrs.
Liefde's Great Dutch admirals.
Life at home, at school, and at college.
Moore's Child's pianoforte book.
Mongan's Our great military commanders.
Müller's (Max) German love.
Reid's (Mayne) The death shot.
,,   The flag of distress.
Taylor (Jeremy), Selections from.
Yonge's Seven heroines of Christendom.
,,   Our great naval commanders.

Gift Books at 3s.

Children's Journey, the.
Gilbert's Modern wonders of the world.
,,   History of a Huguenot Bible.
Keene's Voiceless Teachers.[and Rosette.
Martineau des Chesney's Marquise

O'Reilly's Story of ten thousand homes.
Rowe's Roughing it in Van Diemen's Land.
,,   A haven of rest.
Ross' Dora's boy.
Tytler's Tales of many lands.

Gift Books at 2s. 6d.

Auerbach's Two stories.
Austin's Moon folk.
Bowker's Goblin tales of Lancashire.
Bickerdyke's With the best intentions.
Caballero's Book of Spanish tales.
Camden's Hoity Toity.
,,   Travelling Menagerie.
Cappel's Olu Norse sagas.
Chapman's Notable Women of the
　　Covenant.
Clarke's George and Robert Stephenson.
Cupples' Tappy's chicks.
Fryer's Book of English fairy tales.

Geldart's Modern Greek folk-lore.
Harris' Uncle Remus. *Illust. Edition.*
Howe's The deserted ship.
Irving (W.) Beauties of. 23 plates by G. C.
Kroeker's Alice thro' the looking-glass.
Maitland's Madge Hilton.
Parker's Among the Fairies.
Rowsell's Tales of filial devotion.
,,   Spirit of giant mountains.
Stephens' Old Norse fairy tales.
Thorn's Harlequin eggs; In and Out.
White's Boys of Raby.
Wood's Season among the wild flowers.

Gift Books at 1s. 6d.

Alberg's Charles XII.
Andersen's Fairy tales set to music.
Armstrong's Birds and their ways.
Baker's Stories of olden times.
Bickerdyke's An Irish midsummer night's dream.
Bourne's Life of Gustavus Adolphus.
Chapman's Life of John Wiclif.
    ,,   Life of Martin Luther.
Clarke's Short biographies—

Prince Consort.	George Stephenson.
Charlemagne.	Robert Stephenson.
Sir W. Raleigh.	William Tyndale.

Cobbe's Doll stories.
Cooke's A boy's ideal.
    ,,   True to himself.

Cooke's Latimer's candle.
    ,,   English Hero.
Gausseu's Iron Kingdom.
    ,,   King's Dream.
    ,,   Kingdom of Iron and Clay.
Guizot's Life of Lady Russell.
Hawthorne's Biographical stories.
Richmond's Annals of the poor.
"Robin's" The little flower-girl.
    ,,   Skippo.
Shakspere, his life and times.
Stafford's Only a drop of water.
Stories of my Pets.
Villari's Life in a cave.
Zimmern's Tales from the Edda.

Gift Books at 1s.

Greenwood's Reminiscences of a raven.
Lamb's Mrs. Leicester's school.
Norton's History of Greece for children.
    ,,   History of Rome for children.
    ,,   History of France for children.

Rouse's Sandy's faith.
Sherwood's Juvenile library. 3 vols.
Shield's Knights of the red cross.
Tillotson's Crimson pages.

Gift Books at 6d.

Anson's Cat and dog stories.
    ,,   Three foolish little gnomes.
Chapman's The wanderer.
Cheerful Cherry ; or, make the best of it.
Little Dickie.
    ,,   Goody Two-shoes.
    ,,   Henry and his bearer.
The Babes in a basket.

The Basket of flowers.
The Language and sentiment of flowers.
The Prince in disguise.
The Story of patient Griseldis.
The Perfect home series. By Rev. Dr. Miller. 5 vols., each 6d.
Wallace's Flowers.

MAGAZINES.

TIME. 128 pages, medium 8vo. Monthly. From 1885, 1s.

THE CONTEMPORARY PULPIT. 64 pages, roy. 16mo. Monthly, 6d.

EASTWARD HO! 96 pages. Monthly, 6d.

THE CAMBRIDGE EXAMINER. A Monthly Educational Journal (except July and August). 48 pages, demy 8vo. Monthly, 6d.

THE NATURALIST'S WORLD. Illustrated. 20 pages, fcap. 4to. Monthly, 2d.

THE SCOTTISH NATURALIST. Demy 8vo. Quarterly, 1s. 2d.

Printed by Hazell, Watson, & Viney, Ld., London and Aylesbury.